AF412253

Titles in This Series

THE FIELDS INSTITUTE FOR RESEARCH IN MATHEMATICAL SCIENCES

Hamiltonian and Gradient Flows, Algorithms and Control

Anthony Bloch

Editor

American Mathematical Society
Providence, Rhode Island

The Fields Institute
for Research in Mathematical Sciences

The Fields Institute is named in honour of the Canadian mathematician John Charles Fields (1863–1932). Fields was a remarkable man who received many honours for his scientific work, including election to the Royal Society of Canada in 1909 and to the Royal Society of London in 1913. Among other accomplishments in the service of the international mathematics community, Fields was responsible for establishing the world's most prestigious prize for mathematics research—the Fields Medal.

The Fields Institute for Research in Mathematical Sciences is supported by grants from the Ontario Ministry of Education and Training and the Natural Sciences and Engineering Research Council of Canada. The Institute is sponsored by McMaster University, the University of Toronto, and the University of Waterloo and has affiliated universities in Ontario and across Canada.

1991 *Mathematics Subject Classification.* Primary 34A34, 70H05, 93C10, 65F15, 90C05.

Library of Congress Cataloging-in-Publication Data
Hamiltonian and gradient flows, algorithms and control/Anthony Bloch, editor.
    p. cm.—(Fields Institute communications, ISSN 1069-5265; v. 3)
    ISBN 0-8218-0255-0 (acid-free)
    1. Hamiltonian systems. 2. Control theory. 3. Mathematical optimization. I. Bloch, Anthony, 1955– . II. Series.
QA614.83.H33 1993
$515'.352$—dc20

                                        94-28546
                                              CIP

Contents

Preface

As the title indicates, this volume brings together ideas from several traditionally rather disparate areas of mathematics. The conference at The Fields Institute, which gave rise to these proceedings, was intended to encourage these interactions.

One of the key interactions is that between dynamical systems and algorithms. The classic work of Kostant and Symes showed there was a close link between the QR algorithm and the Toda lattice flow in the form derived by Flaschka and generalized by Bogoyavlenskij, Kostant and others. The classical Toda lattice flow describes the dynamics of a set of 1-dimensional particles interacting under an exponential potential and is an integrable Hamiltonian system. Flaschka showed it could be written in Lax pair form. David Watkins pointed out during the conference that the Toda lattice equations in Kostant's variables actually appeared in Rutishauser's work on the QR algorithm, quite independently of any connection with Hamiltonian flows or lattices. The links between Toda and the QR algorithm were considerably strengthened in the work of Deift, Li, Nanda and Tomei.

Another link between algorithms and dynamics was forged in the context of interior point methods for linear programming and Karmarkar's algorithm. Bayer and Lagarias analyzed the geometry of Karmarkar's algorithm for interior point programming in the context of smooth dynamical systems, and they showed that certain smooth flows associated with it are integrable Hamiltonian systems.

A major catalyst for some of the current work linking dynamics and algorithms has been the innovative recent work of Brockett. Brockett introduced a smooth flow (the double bracket equation) which carried out a number of discrete calculations including the sorting of numbers and certain linear programming problems. Bloch, Brockett, Flaschka and Ratiu studied the link between this equation and the Toda lattice flow, as well as the links with the theory of the convexity of the moment map of Schur-Horn-Kostant-Atiyah-Guillemin-Sternberg. Further applications to linear programming have been given by Faybusovich. Brockett has also shown that all finite automata may be encoded in a related dynamical systems form. Fundamental related work in this regard includes the work of Blum and Smale and their collaborators on automata over the reals. This work is a deep study of how computation may be extended to the continuous domain. The work of Chu has described the links between a number of algorithms and flows, while Helmke and Moore have used dynamics to analyze a number of optimization problems related to control theory. Some of these ideas are also discussed in the recently published book by Helmke and Moore.

This brings us to two other connections indicated in the title of this volume: firstly that between integrable Hamiltonian and gradient flows, and secondly that between Hamiltonian and gradient dynamics and control. Regarding the former, it was shown, following the work of Moser, that the Toda lattice flow has a dual

Hamiltonian and gradient character, the key to its connection with algorithms. This duality also plays a major role in the work of Bayer and Lagarias. Regarding the latter, optimal control has, of course, a classic relationship with Hamiltonian systems, but one which has been strengthened in a number of interesting ways recently. For example, in the paper by Brockett in this volume, it is shown that there is an interesting optimal control problem which gives rise to the Toda lattice equations, as well as an optimal control problem which gives rise to the general double bracket equation. In the paper by Bloch, Crouch and Ratiu, optimal control problems which give rise to singular generalized rigid body equations are discussed.

The paper by Smith shows how geometry and dynamics can be brought to bear on the problem of solving optimization problems on Riemannian manifolds. In this context, new versions of the conjugate gradient method and Newton's method are developed. In the paper by Faybusovich, linear programming, dynamical systems and the Gibbs variational principle of statistical physics are linked. Wong uses gradient flows to study the assignment problem, while the paper by Chu describes a number of matrix flows with applications to numerical problems. Ammar and Gragg relate flows on Hessenberg matrices to eigenvalue computations. The paper by Sturmfels estimates the asymptotic number of real roots of a system of polynomial equations and relates this computation to the Newton polytope and the work of Koushnirenko. This is, in turn, closely related to the moment map discussed above.

The papers of Alber and Marsden and of Bogoyavlenskij extend the territory covered here to PDE's. Alber and Marsden show how geometric phases occur in soliton equations and use them to analyze the geometry and dynamics of these equations. Bogoyavlenskij analyzes systems of hydrodynamic type, which are natural infinite-dimensional extensions of the finite Toda lattice and Volterra models. (In fact the extension of a number of the ideas discussed in this volume to PDE's is proving to be a very rich area of research, from both a dynamic and geometric point of view.) Finally, Brunei illustrates the geometry of isospectral deformations of matrices on the connection machine.

These papers provide a glimpse of the work that has been done in this fascinating intersection of fields. There is much ongoing research and I believe that many more exciting developments are still to come.

I would like to thank Jerry Marsden, President and Scientific Director of The Fields Institute, Bill Shadwick, Vice-President and Executive Director of The Fields Institute, Robin Skinner, who was invaluable in producing this volume, Judy Motts and the rest of the staff who helped to arrange the workshop, and all the participants of the workshop, for making the workshop and these proceedings possible.

Anthony Bloch

Fields Institute Communications
Volume **3**, 1994

Resonant Geometric Phases for Soliton Equations

M.S. Alber
Department of Mathematics
University of Notre Dame, Notre Dame
IN 46556, USA

J.E. Marsden
Department of Mathematics
University of California, Berkeley
CA 94720, USA

To the memory of Solomon J. Alber

1 Introduction

Methods of symplectic and complex geometry have recently given important information about the phase-space geometry of nonlinear equations (see, for example, Ercolani, Forest, McLaughlin, and Montgomery [1987], Ercolani [1989], Ercolani and McLaughlin [1991], McLaughlin and Overman [1993]). A few specific examples are as follows. From the point of view of complex geometry, the reality conditions and homoclinic varieties for the sine-Gordon (SG) equation

$$U_{xx} - U_{tt} = \sin U, \tag{1.1}$$

were investigated in Ercolani and Forest [1985] and Ercolani [1989]. The modulational Poisson structure for the sine-Gordon system was derived in terms of conformal ingredients, such as differentials on Riemann surfaces and θ-functions, and a link with the Hamiltonian theory was investigated (see Ercolani et. al. [1987]). Geometric singular perturbation theory was discussed in Kovacic and Wiggins [1992] in connection with the resonances of the sine-Gordon equation. In Bishop, McLaughlin and Solerno [1989] a global 2-dimensional system of real action-angle variables was constructed for the sine-Gordon equation, which gives a framework for describing the breather-kink-antikink transition. Then a geometric method of "blowing-up the singularity" was used for completing the phase space in the neighborhood of the singular point of the corresponding dynamical system and for extending it to a Hamiltonian system on this new phase space. (For details about a method of "blowing-up the singularity", see Birnir [1986].) Complex geometry related to

---

1991 *Mathematics Subject Classification.* Primary 58F07; Secondary 70H99.

Supported in part by the Ministry of Colleges and Universities of Ontario and the Natural Sciences and Engineering Research Council of Canada while visiting The Fields Institute.

quasiperiodic and soliton solutions and singular theta functions was studied in the KdV case in McKean [1977] and Ercolani [1989]. Finally, a topological classification of soliton equations based on the geometry of compact invariant varieties was studied in Ercolani and McLaughlin [1991].

In Alber and Marsden [1992, 1994] we developed a new method for obtaining geometric phase phenomena for soliton equations, including the familiar phase shift of interacting solitons. Traditionally, the phase spaces of integrable systems were viewed as being foliated by invariant tori; however, for soliton phase spaces and to get formulas for geometric phases, we have shown that a foliation by *noncompact* varieties is essential. The method is based on the introduction of a new complex angle representation on a noncompact invariant variety. (In particular, a 1-soliton solution is identified with the linearized Hamiltonian flow on a logarithmic Riemann surface.) Asymptotic reduction of the n-dimensional soliton complex angle representation then yields a description of geometric phases in terms of the monodromy at singularities in the space of parameters.

In the process of investigating these angle representations, we deal in a natural way with the phenomena of geometric phases. Recall that in Berry [1984], a geometric phase factor $\exp(i\gamma)$ was considered for systems that are slowly (adiabatically) transported along closed curves in a space of parameters. In Montgomery [1988] a class of connections was constructed to obtain expressions for the Hannay-Berry phases (these are geometric angle shifts in the classical case) for some integrable problems in terms of the nontrivial holonomy of these connections. (See Berry and Hannay [1988].) Montgomery also gave an example of a phase associated with the presence of singularities in the case of a flat connection. Symmetry and reduction were used in Marsden, Montgomery and Ratiu [1990] to obtain a generalization of geometric phases to the non-integrable case in the form of the holonomy of the Cartan-Hannay-Berry connection. Recently, David and Holm [1992] investigated Hannay-Berry phases for the real Maxwell-Bloch equations on $\mathbf{R}^3$.

As was already discussed, we introduced angle representations on noncompact Jacobi varieties and used asymptotic reduction to obtain geometric phases for a pair of commuting soliton Hamiltonian flows. Our method uses the fact that asymptotic reduction leads to the complex splitting of the spectrum of the soliton problem. A connection between the θ-function and the τ-function can also be used to find a link between soliton geometric phases and geometric phases for quasiperiodic solutions (see Alber and Marsden [1992] and Alber [1992]). We note that in the quasiperiodic case, geometric phases are related to the Whitham equations through the phase function.

The goal of the present paper is to introduce a multidimensional generalization of asymptotic reduction given in Alber and Marsden [1992], to use this to obtain a new class of solutions that we call resonant solitons, and to study the corresponding geometric phases. The term "resonant solitons" is used because these solutions correspond to a spectrum with multiple points, and they also represent a dividing solution between two different types of solitons. In this sense, these new solutions are degenerate and, as such, will be considered as singular points in the moduli space of solitons.

As one passes from n-soliton solutions to resonant solitons, the spectrum changes continuously, but the invariant variety on which the solitons reside, does not. Thus, one has to use the tools of any perturbative set up with great care. The methods of asymptotic reduction provide these tools.

These methods give a detailed description of complex phases for the focussing nonlinear Schrödinger, or (f)NLS, equation,

$$i\dot{Q} + \frac{1}{2}Q'' + \bar{Q}Q^2 = 0 \tag{1.2}$$

and the sine-Gordon equation, including phases for the breather-kink-antikink and soliton-separatrix interactions. They also yield a link with the geometric models for resonant solutions, such as umbilic geodesics on quadrics. (See Alber and Marsden [1994].) Our approach also provides a link between solitons and homoclinic orbits of nonlinear systems and should enable one to investigate perturbations of the angle representations of nonlinear systems to detect chaotic behavior in the neighborhood of the homoclinic orbits. In particular, in Alber and Marsden [1994] we show that the Hamiltonian flow associated with homoclinic orbits introduced in Devaney [1978] for the C. Neumann problem coincides with the soliton x-flow of the KdV equation. This result leads us naturally to the introduction of homoclinic geometric phases.

We summarize some of the new features found in the present paper using generalizations of the method of asymptotic reduction:

- Our general approach allows one to study the collision, or interaction, of solitons of different type, such as the interaction of a breather and a kink solution, which generates additional phase shifts. The paper also discusses resonant solutions, which, as we have mentioned are degenerate solitons, such as solutions separating breathers and kinks.

- We show that along a 2-dimensional direction in an invariant variety V, there is a splitting of the $2n$-dimensional resonant angle representation into the sum of n copies of 2-dimensional solutions of the same type. This type of behavior was previously observed in the case of soliton solutions along a 1-dimensional direction. Namely, this splitting is the analogue, for resonant solutions, of the familiar asymptotic splitting of an n-soliton solution into n individual 1-soliton solutions.

- Finally, we develop the method of *discrete* asymptotic reduction and demonstrate it using the Toda lattice and new discretizations of the defocusing nonlinear Schrödinger (d)NLS equation and obtain the corresponding phases.

We start by recalling in Section 2 the 1-dimensional method of asymptotic reduction for the associated complex angle representations that yields, among other things, complex geometric phases. In Section 3 we illustrate this method using 2-soliton solutions of the focusing (f)NLS equation. We introduce a new kind of complex soliton generated by a collision of two standard solitons, which we call a "ghost" soliton. The word "ghost" is used because one "sees" this part of the solution in the complexification of the system, which perhaps explains why the object was missed up to now. This ghost soliton is necessary for the understanding of the entire interaction and for the asymptotic splitting of the solitons. We also expect that it will play an important role in the perturbation theory of the (f)NLS equation and in applications to nonlinear optics, especially in view of recent efforts to use optical solitons in fiber optics and the importance of temporal and spatial phase shifts in these investigations. (See, for example, Hasegawa and Kodama [1992].)

The general multi-dimensional method of asymptotic reduction is described in Section 4. This approach is illustrated for the case of resonant solutions of nonlinear equations and is applied to the collision, or interaction, of solitons of different type, such as the interaction of a breather with a kink solution of the sine-Gordon equation as well as the interaction of a soliton with a separatrix. In particular, as is shown in Section 5, such interactions generate new phase shifts, including resonant geometric phases.

Finally, in Section 6 we describe asymptotic reduction for discrete (infinite particle) nonlinear problems that arise from partial differential equations by an "exact" discretization process. We will show that this includes Toda lattices that arise from discretizing the soliton solutions of the defocussing nonlinear Schrödinger equation. (This discretization process is different from standard methods of approximating nonlinear equations described in Toda [1989]; here we *sample* the *solutions* of nonlinear equations at discrete spatial points and get an exact solution of the infinite particle Toda lattice equations.) This leads to an introduction of new exponential Hamiltonians, angle representations and geometric phases.

The general theory of continuous and discrete systems that are related to each other through the same Jacobi varieties (level sets in the phase space) was developed by Solomon J. Alber [1989, 1991].

Recently methods of asymptotic reduction were applied to the study of peaked solitons and billiard solutions of a class of nonlinear integrable pde's in Alber, Camassa, Holm and Marsden [1994a]. This class of equations includes a shallow water equation investigated in Camassa and Holm [1993] and Camassa, Holm and Hyman [1993] as well as equations from the Dym hierarchy. The method we use for the shallow water equation also leads to a link between a certain class of solutions of one of the members of the Dym hierarchy and solutions of the Hamiltonian system for an N-dimensional elliptic billiard. The details of this link can be found in Alber, Camassa, Holm and Marsden [1994b].

2 Complex geometric phases

In Alber and Marsden [1992, 1994] a general approach to geometric phases for soliton equations was described and the (d)NLS and KdV equations were considered in detail.

In what follows, we demonstrate the general method of asymptotic reduction of the complex angle representations for the (f)NLS and sine-Gordon (SG) hierarchies of equations and use methods of complex analysis to yield information about geometric phases. In particular, this method gives a description of a new effect of a residual interaction between 1-soliton solutions after the asymptotic splitting of the n-soliton solution of the (f)NLS equation. On the other hand, angle representations for the SG equation give insight into the breather-kink-antikink interaction and describe the effect of such interactions on the transition through the separatrix solution.

One starts with two complex Hamiltonian flows in complex phase space $\mathbf{C}^{2n}$. The main purpose of asymptotic reduction is to investigate the vector phase function φ, a $\mathbf{C}^n$ valued meromorphic function defined on an invariant variety V consisting of the symmetric product of n copies of a Riemann surface $\Re \subset \mathbf{C}^{2n}$ associated to the problem. This Riemann surface $\Re$ is determined by the constants of the motion of the particular problem, and was defined for different examples in Alber and

Marsden [1992]. Asymptotic reduction plays the role of averaging in the sense that it separates the geometric phase from the dynamical phase. Normally the dynamic phase is the frequency times the time (or the integral of the frequency with time (see, for example, Montgomery [1988], page 276). In the case of solitons, the period is infinite, and this is a main difficulty that is overcome by the method of asymptotic reduction. The complex phase function, which stands as an argument in the τ-function, keeps track of the phase information in solitons, even when they are interacting. In this context, the phase shift of solitons is related to the singularity structure of the phase function.

We may regard V as a bundle over a space of parameters M. In our case, the parameters consist of spectral parameters a_j that will be discussed below. We can choose an arbitrary connection on this bundle. In what follows, we will choose a trivial connection because we are primarily interested in describing geometric phases in terms of monodromy caused by the presence of singularities. In this context, the geometric phase is defined, as in Alber and Marsden [1992], as follows.

Definition 2.1 *We define* **soliton geometric phases** *as:*

$$\oint_C d_a \varphi. \tag{2.1}$$

Here d_a denotes the covariant derivative relative to the given connection and C is a closed curve in the space of parameters.

Note that this formula incorporates the possibilities of two sources for geometric phases, namely the anholonomy of a connection, and the monodromy of a singularity. From the point of view of a connection it is a natural extension of the results obtained for the Cartan-Hannay-Berry connection in the real case using averaging techniques

$$\text{Hannay's angles} = \oint_C \langle d_a \theta \rangle . \tag{2.2}$$

For details see Montgomery [1988] and Marsden, Montgomery and Ratiu [1990].

3 Geometric phases for the (f)NLS equation

It is known that quasiperiodic invariant tori are degenerate for the NLS equation. These degeneracies are related to the fact that the NLS equation should be considered as a degenerate case of a more general nonlinear problem. (For details see Previato [1985] and M. Alber and S. Alber [1985, 1987].)

This difficulty manifests itself also on the level of searching for solitons. In what follows, we show that the method of asymptotic reduction naturally deals with the degenerate nature of the soliton solutions on the level of invariant varieties.

To obtain multi-solitons of the (f)NLS equation, we first consider quasiperiodic solutions. We start with the spectral polynomial

$$C(\mu) = \prod_{r=1}^{4N} (\mu - m_r) \tag{3.1}$$

depending parametrically on the spectrum $m_1, \ldots, m_{4N}$. The complex numbers m_j are assumed to be distinct. Let

$$D_j = \left(- \sum_{l=1, l \neq j}^{g} \mu_l + \frac{1}{2} \sum_{r=1}^{4N} m_r \right). \tag{3.2}$$

Here $g = 2N - 1$. Next, consider the following Riemann surface:

$$\Re_{quasi} : \quad W^2 = -C(\mu) \tag{3.3}$$

and the variety $V = (\Re_{quasi} \times \cdots \times \Re_{quasi})/\sigma_g$ where σ_g is the group of permutations of g letters. Finally, let $\epsilon_j = \pm 1$. One can choose these arbitrarily at the moment; their significance will be discussed later. In terms of this data, we consider the following two systems of differential equations on V:

$$\left. \begin{aligned} \mu'_j &= \frac{\partial \mu_j}{\partial x} = 2i\epsilon_j \frac{\sqrt{C(\mu_j)}}{\prod_{l \neq j}(\mu_j - \mu_l)} \\[2ex] \dot{\mu}_j &= \frac{\partial \mu_j}{\partial t} = 2i\epsilon_j D_j \frac{\sqrt{C(\mu_j)}}{\prod_{l \neq j}(\mu_j - \mu_l)}, \quad j = 1, ..., g, \end{aligned} \right\} . \tag{3.4}$$

Here each variable μ_j lies on a copy of the Riemann surface $\Re_{quasi}$.

Each of these equations has first integrals given by:

$$P_j^2 = -C(\mu_j) = - \prod_{r=1}^{4N}(\mu_j - m_r), \quad j = 1, ..., g. \tag{3.5}$$

The equations may be considered as a Hamiltonian system with conjugate variables μ_j and P_j.

Now we apply the following limiting process to the spectrum m_r of the quasiperiodic problem:

$$m_{4k}, m_{4k-1} \to a_k, \quad m_{4k-2}, m_{4k-3} \to \bar{a}_k, \quad k = 1, ..., N. \tag{3.6}$$

Here $N = (g+1)/2$ and g, the genus of $\Re_{quasi}$, is assumed to be an odd number. (See Previato [1985] and McKean [1977] for details.) This limiting process defines a new polynomial $\tilde{C}$ by $\sqrt{C(\mu)} \to \tilde{C}(\mu)$ and results in the following limiting differential equations (3.4):

$$\left. \begin{aligned} \frac{\partial \mu_j}{\partial x} &= 2i\epsilon_j \frac{\prod_{r=1}^{N}(\mu_j - a_r)(\mu_j - \bar{a}_r)}{\prod_{l \neq j}(\mu_j - \mu_l)}, \\[2ex] \frac{\partial \mu_j}{\partial t} &= 2i\epsilon_j D_j \frac{\prod_{r=1}^{N}(\mu_j - a_r)(\mu_j - \bar{a}_r)}{\prod_{l \neq j}(\mu_j - \mu_l)}, \quad j = 1, ..., g \end{aligned} \right\} \tag{3.7}$$

describing an N-soliton solution of the (f)NLS equation. Here

$$D_j = \left(- \sum_{l=1, l \neq j}^{g} \mu_l + \sum_{k=1}^{N}(a_k + \bar{a}_k) \right),$$

and ϵ_j is as before. Note that prior to Alber and Marsden [1992], the limiting process (3.6) was usually applied to the system of first integrals as well. As a result, the soliton Hamiltonian flows were considered on pinched tori in the phase space. We have shown that, instead, by introducing the new system of first integrals

$$P_j = \sum_{k=1}^{N} \log(\mu_j - a_k)(\mu_j - \bar{a}_k) \quad j = 1, ..., g, \tag{3.8}$$

the system can be realized on a noncompact invariant variety: $(\Re_{sol} \times \cdots \times \Re_{sol})/\sigma_g$, where the Riemann surface $\Re_{sol}$ is defined by

$$\Re_{sol} : \ W = \sum_{k=1}^{N} \log(\mu - a_k)(\mu - \bar{a}_k). \tag{3.9}$$

In Alber and Marsden [1992] we showed that the system (3.7) together with its new first integrals (3.8) can be realized as a Hamiltonian system with an exponential complex Hamiltonian.

Note that the limiting process described above yields an odd dimensional complex (f)NLS system. In what follows we will demonstrate that this yields an additional complex soliton generated by a collision of the standard solitons.

3.1 2-dimensional soliton solution of (f)NLS equation Before describing the N-soliton solutions, we want to give a motivation for our approach by describing the "ghost" soliton generated by a collision of two standard optical solitons.

An analysis of the Riemann surface (3.9) associated with the 2-soliton solution of the (f)NLS equation yields 3 μ-variables (μ_1, μ_2, μ_3). This gives 3 terms in the angle representation $(\theta_1, \theta_2, \theta_3)$ and demonstrates the existence of the 3-rd "ghost" soliton. In what follows we use the notation: $a_1 = x_1 + iy_1$ and $a_2 = x_2 + iy_2$.

The angle representation that describes the collision of 2 standard solitons of the (f)NLS equation, namely

$$
\begin{aligned}
\theta_1 &= -\frac{1}{4y_1} \left(\sum_{j=1}^{2} \int_{\mu_j^0}^{\mu_j} \left(\frac{1}{(\mu_j - a_1)} - \frac{1}{(\mu_j - \bar{a}_1)} \right) d\mu_j \right) \\
&\quad + \frac{1}{4y_1} \left(\int_{\mu_3^0}^{\mu_3} \left(\frac{1}{(\mu_3 - a_1)} - \frac{1}{(\mu_3 - \bar{a}_1)} \right) d\mu_3 \right) \\
&= x + v_1 t,
\end{aligned}
\tag{3.10}
$$

where $v_1 = 2x_1$ and

$$
\begin{aligned}
\theta_2 &= -\frac{1}{4y_2} \left(\sum_{j=1}^{2} \int_{\mu_j^0}^{\mu_j} \left(\frac{1}{(\mu_j - a_2)} - \frac{1}{(\mu_j - \bar{a}_2)} \right) d\mu_j \right) \\
&\quad + \frac{1}{4y_2} \left(\int_{\mu_3^0}^{\mu_3} \left(\frac{1}{(\mu_3 - a_2)} - \frac{1}{(\mu_3 - \bar{a}_2)} \right) d\mu_3 \right) \\
&= x + v_2 t,
\end{aligned}
\tag{3.11}
$$

where $v_2 = 2x_2$, also includes the 3-rd angle variable, which accounts for the remaining interaction between two initial solitons after asymptotic splitting. The third angle variable yields an additional soliton and can be chosen for $t \to \infty$ and $t \to -\infty$ as follows

$$
\theta_3^+ = -\frac{1}{2(y_1 - y_2)} \left(\sum_{j=1}^{2} \int_{\mu_j^0}^{\mu_j} \left(\frac{1}{(\mu_j - a_1)} - \frac{1}{(\mu_j - a_2)} \right) d\mu_j \right)
$$

$$
+ \frac{1}{2(y_1 - y_2)} \left(\int_{\mu_3^0}^{\mu_3} \left(\frac{1}{(\mu_3 - a_1)} - \frac{1}{(\mu_3 - a_2)} \right) d\mu_3 \right) = (x + v_3 t) + \mathrm{Im}(\theta_3),
$$

$$
v_3 = \frac{2(y_1 x_1 - y_2 x_2)}{(y_1 - y_2)}
\tag{3.12}
$$

and

$$
\theta_3^- = \frac{1}{2(y_1 - y_2)} \left(\sum_{j=1}^{2} \int_{\mu_j^0}^{\mu_j} \left(\frac{1}{(\mu_j - \bar{a}_1)} - \frac{1}{(\mu_j - \bar{a}_2)} \right) d\mu_j \right)
$$

$$
- \frac{1}{2(y_1 - y_2)} \left(\int_{\mu_3^0}^{\mu_3} \left(\frac{1}{(\mu_3 - \bar{a}_1)} - \frac{1}{(\mu_3 - \bar{a}_2)} \right) d\mu_3 \right) = (x + v_3 t) + \mathrm{Im}(\theta_3),
$$

$$
v_3 = \frac{2(y_1 x_1 - y_2 x_2)}{(y_1 - y_2)}.
\tag{3.13}
$$

Here $\theta_3^- = -\bar{\theta}_3^+$. To illustrate the phase shift procedure, we assume that $(x_2 < x_1)$ and $(0 < y_2 < y_1)$, which implies that $(v_2 < v_1)$. Asymptotic reduction along $(d_1 = x + v_1 t)$ and $(d_2 = x + v_2 t)$ (described in Alber and Marsden [1992]) yields a splitting of the 2-soliton angle representation and results in the following phase shifts:

$$
\left. \begin{aligned}
|\Delta_G \theta_1| &:= \frac{1}{y_1} \log \left| \frac{a_2 - a_1}{a_2 - \bar{a}_1} \right|, \\[2mm]
Arg \Delta_G \theta_1 &:= 2 \arctan \left(\frac{y_2 - y_1}{x_2 - x_1} \right) - 2 \arctan \left(\frac{y_2 + y_1}{x_2 - x_1} \right) \\[2mm]
|\Delta_G \theta_2| &:= -\frac{1}{y_2} \log \left| \frac{a_1 - a_2}{a_1 - \bar{a}_2} \right|, \\[2mm]
Arg \Delta_G \theta_2 &:= -2 \arctan \left(\frac{y_2 - y_1}{x_2 - x_1} \right) + 2 \arctan \left(\frac{y_2 + y_1}{x_1 - x_2} \right).
\end{aligned} \right\}
\tag{3.14}
$$

Reduction along the third real direction $(d_3 = x + v_3 t)$ yields an additional phase shift, namely the phase shift of the "ghost" soliton corresponding to θ_3:

$$
|\Delta_G \theta_3| = \frac{1}{(y_1 - y_2)} \log \left| \frac{y_1}{y_2} \right|.
\tag{3.15}
$$

We choose $\theta_3 = \theta_3^+$ as $t \to \infty$ and $\theta_3 = \theta_3^-$ as $t \to -\infty$.

We believe that these new soliton phase shifts can be detected in an experiment similar to those used to measure temporal phase shifts of the optical solitons (see Friberg, Machida and Yamamoto [1992] and Hasegawa and Kodama [1992]).

Remark 3.1 Note that the condition $v_1 = v_2 = v$ (or $x_1 = x_2$) yields $v_3 = v$. In other words, the additional soliton generated by standard solitons moving with identical speeds, does not split from them.

To clarify the link between complex angle representations introduced above and standard formulae for optical solitons from nonlinear optics (see Ablowitz and Segur [1981] and Hasegawa and Kodama [1992]) we describe below the 1-dimensional case; that is, the case ($N = 1, g = 2N - 1 = 1$). There is only one $\mu(x,t)$ variable which satisfies the following system of differential equations

$$\left.\begin{aligned} \frac{\partial \mu}{\partial x} &= 2i\varepsilon(\mu - a)(\mu - \bar{a}) \\[2ex] \frac{\partial \mu}{\partial t} &= 2i\varepsilon(\mu - a)(\mu - \bar{a})D. \end{aligned}\right\} \tag{3.16}$$

Here $D = (a + \bar{a}) = 2\mathrm{Re}(a)$. In what follows we consider $\varepsilon = 1$. The choice of the basic point of the angle representation on a particular sheet of the Riemann surface plays an important role in asymptotic reduction for complex solutions of the (f)NLS equation.

The system (3.16) yields:

$$-\frac{1}{2i} \int \frac{d\mu}{(a - \mu)(\mu - \bar{a})} = x + vt, \quad v = 2\mathrm{Re}(a), \tag{3.17}$$

or

$$\frac{1}{4\mathrm{Im}(a)} \int \left(\frac{1}{(a - \mu)} + \frac{1}{(\mu - \bar{a})}\right) d\mu = x + vt. \tag{3.18}$$

From the equivalent expression

$$\log\left(\frac{\mu - \bar{a}}{a - \mu}\right) = 4\mathrm{Im}(a)(x + vt + \theta_o) = \beta \tag{3.19}$$

it follows that

$$\frac{\mu - \bar{a}}{a - \mu} = e^{\beta}, \quad \mu = \frac{\bar{a} + ae^{\beta}}{1 + e^{\beta}} = \mathrm{Re}(a) - i\mathrm{Im}(a)\left(\frac{1 - e^{\beta}}{1 + e^{\beta}}\right). \tag{3.20}$$

Now, the solution $Q(x,t)$ of the (f)NLS equation is related to the μ-variable $\mu(x,t)$ by

$$U = -2\mu + 2(a + \bar{a}) = -2\mu + 4\mathrm{Re}(a) = i\frac{\partial}{\partial x}\log Q, \tag{3.21}$$

and therefore, we have

$$Q = \exp\left(-i\int U dx\right) = \exp\left(2i\int \mu dx - 4i\mathrm{Re}(a)x\right). \tag{3.22}$$

Here

$$\int \mu \, dx = \mathrm{Re}(a)x - i\mathrm{Im}(a) \int \left(\frac{1 - e^{\beta}}{1 + e^{\beta}}\right) dx \tag{3.23}$$

and

$$\begin{aligned}
\int \left(\frac{1 - e^{\beta}}{1 + e^{\beta}}\right) dx &= \int \frac{dx}{1 + e^{\beta}} - \int \frac{e^{\beta} dx}{1 + e^{\beta}} = \int \frac{e^{-\beta} dx}{e^{-\beta} + 1} - \int \frac{e^{\beta} dx}{1 + e^{\beta}} \\
&= -\frac{1}{4\mathrm{Im}(a)} \log(e^{-\beta} + 1) - \frac{1}{4\mathrm{Im}(a)} \log(1 + e^{\beta}) + rt + c_0 \\
&= -\frac{1}{4\mathrm{Im}(a)} \log \left(\frac{e^{-\frac{\beta}{2}} + e^{\frac{\beta}{2}}}{2}\right)^2 + \delta t + C_0.
\end{aligned} \tag{3.24}$$

Finally, we obtain the expression

$$Q = Q^0 \exp\left(-2i\mathrm{Re}(a)x + 2i\mathrm{Im}(a)\delta t - i\nu_0\right) \mathrm{sech}\left(2\mathrm{Im}(a)(x + vt + \theta_0)\right).$$

After substituting the expression for Q into equation (1.2) we determine the coefficient δ. This results in the formula

$$lQ = Q^0 \exp\left(-2i\mathrm{Re}(a)x + 2i((\mathrm{Im}(a))^2 - (\mathrm{Re}(a))^2)t - i\nu_0\right) \mathrm{sech}\left(2\mathrm{Im}(a)(x + vt + \theta_0)\right)$$

 which coincides with the expression used in nonlinear optics (see Hasegawa and Kodama [1992]).

We expect that angle representation and new phase shifts described above will play an important role in applications to nonlinear optics, especially in view of recent efforts to use optical solitons in fiber optics and the importance of temporal and spatial phase shifts in these investigations. In particular, angle representations can be used for numerical simulations of the solutions of perturbed nonlinear equations.

3.2 n-soliton solutions of the (f)NLS equation First of all, we show that (3.7) gives rise to a new Hamiltonian system in the phase space $\mathbf{C}^{2g}$.

Theorem 3.2 *The systems (3.7) are soliton Hamiltonian systems with Hamiltonians*

$$H_s^s = 2i \frac{\sum_{j=1}^{g} (e^{2iP_j} - \bar{C}(\mu_j))}{\prod_{l \neq j}(\mu_j - \mu_l)} \tag{3.25}$$

and

$$H_s^d = 2i \frac{\sum_{j=1}^{g} D_j (e^{2iP_j} - \bar{C}(\mu_j))}{\prod_{l \neq j}(\mu_j - \mu_l)} \tag{3.26}$$

and with first integrals

$$P_j = \frac{\epsilon_j}{2i} \sum_{r=1}^{N} \log(\mu_j - a_r)(\mu_j - \bar{a}_r) \quad j = 1, ..., g, \tag{3.27}$$

which determine an invariant variety in the form of a symmetric product

$$(\Re_{sol} \times \cdots \times \Re_{sol})/\sigma_g,$$

of n copies of the Riemann surface described as follows

$$\Re_{sol} : \; W = \frac{1}{2i} \sum_{r=1}^{N} \log(\mu - a_r)(\mu - \bar{a}_r). \tag{3.28}$$

Theorem 3.3 *The Hamiltonian systems (3.25) and (3.26) have a system of action-angle variables which splits into N 1-soliton angle variables with corresponding phase functions as $(t \to \infty)$ or $(t \to -\infty)$.*

Proof
From the system (3.7), we obtain the following system:

$$\left.\begin{aligned}
\frac{\mu_j'}{(\mu_j - a_r)(\mu_j - \bar{a}_r)} &= \epsilon_{rj} \frac{\prod_{s \neq r}^{N}(\mu_j - a_s)(\mu_j - \bar{a}_s)}{\prod_{l \neq j}(\mu_j - \mu_l)}, \\[2ex]
\frac{\dot{\mu}_j}{(\mu_j - a_r)(\mu_j - \bar{a}_r)} &= \epsilon_{rj} D_j \frac{\prod_{s \neq r}^{N}(\mu_j - a_s)(\mu_j - \bar{a}_s)}{\prod_{l \neq j}(\mu_j - \mu_l)}, \quad r = 1, ..., N.
\end{aligned}\right\} \tag{3.29}$$

We take into consideration parameters $\epsilon_{rj} = \pm 1$ inherited from the quasiperiodic problem. They will determine in what follows the choice of the sheet of the Riemann surface in the expression for the r-th angle variable. Note that a soliton solution corresponds to a particular choice of the indices. Otherwise, the system (3.29) describes a more general class of solutions.

Summing the above equations (3.29) with respect to j, we obtain the first part of the angle representation

$$\begin{aligned}
\theta_r &= -\frac{1}{4\mathrm{Im}(a_r)} \sum_{j=1}^{N} \int_{\mu_j^o}^{\mu_j} \left(\frac{1}{(\mu_j - a_r)} - \frac{1}{(\mu_j - \bar{a}_r)} \right) d\mu_j \\[2ex]
&\quad + \frac{1}{4\mathrm{Im}(a_r)} \sum_{j=1, j \neq k}^{N} \int_{\mu_j^o}^{\mu_j} \left(\frac{1}{(\mu_{j+N} - a_r)} - \frac{1}{(\mu_{j+N} - \bar{a}_r)} \right) d\mu_{j+N} \\[2ex]
&= x + v_r t, \tag{3.30}
\end{aligned}$$

where $v_r = 2\mathrm{Re}(a_r)$ and $r = 1, ..., N$. Here $1 \leq k \leq N$.

We will consider the real part of the angle representation first. It will be shown in the end of this section that phases for the imaginary part of the angle representation are determined by the phases for the real part.

Now we fix the k-th direction, $d_k = x + v_k t =$ constant, and investigate the behavior of the rest of the angle variables

$$\theta_r = x + (v_r - v_k)t + v_k t = d_k + (v_r - v_k)t, \; r = 1, ..., N : \; r \neq k, \tag{3.31}$$

as $t \to \pm\infty$. This yields the following limits for the first $(N-1)$ μ_r-variables:

$$\left.\begin{aligned}
&1. \; t \to \infty \; : \; \mu_r \to a_r, \text{ for } r < k \text{ and } \mu_r \to \bar{a}_r \text{ for } N \geq r > k \\[2ex]
&2. \; t \to -\infty \; : \; \mu_r \to \bar{a}_r \text{ for } r < k \; \text{ and } \; \mu_r \to a_r \text{ for } N \geq r > k.
\end{aligned}\right\} \tag{3.32}$$

The important part of asymptotic reduction in the case of (f)NLS equation is that it should be applied to the whole vector space of angle representations which has a basis consisting of $g = (2N - 1)$ elements. We choose ϵ_{jr} in a way described below and determine $(N-1)$ missing elements of the basis in each of the cases $t \to \infty$ and $t \to -\infty$, as follows:

1. $(t \to \infty)$ for $r < k$, let

$$
\begin{aligned}
\theta_{r+N} &= \sum_{j=1}^{N} \int_{\mu_j^o}^{\mu_j} \left(\frac{1}{(\mu_j - a_k)} - \frac{1}{(\mu_j - a_r)} \right) d\mu_j \\
&\quad + \sum_{j=1,j\neq k}^{N} \int_{\mu_{j+N}^o}^{\mu_{j+N}} \left(\frac{1}{(\mu_{j+N} - a_k)} - \frac{1}{(\mu_{j+N} - a_r)} \right) d\mu_{j+N} \\
&= 2i(a_k - a_r)(x + v_{r+N}t),
\end{aligned}
\tag{3.33}
$$

and for $N \geq r > k$, let

$$
\begin{aligned}
\theta_{r+N} &= \sum_{j=1}^{N} \int_{\mu_j^o}^{\mu_j} \left(\frac{1}{(\mu_j - \bar{a}_k)} - \frac{1}{(\mu_j - \bar{a}_r)} \right) d\mu_j \\
&\quad - \sum_{j=1,j\neq k}^{N} \int_{\mu_{j+N}^o}^{\mu_{j+N}} \left(\frac{1}{(\mu_{j+N} - \bar{a}_k)} - \frac{1}{(\mu_{j+N} - \bar{a}_r)} \right) d\mu_{j+N} \\
&= 2i(\bar{a}_k - \bar{a}_r)(x + v_{r+N}t).
\end{aligned}
\tag{3.34}
$$

Here

$$
\left.
\begin{aligned}
v_{r+N} &= a_k + a_r \quad \text{for} \quad r < k, \\
v_{r+N} &= \bar{a}_k + \bar{a}_r \quad \text{for} \quad r > k.
\end{aligned}
\right\}
\tag{3.35}
$$

2. $(t \to -\infty)$ for $r < k$, let

$$
\begin{aligned}
\theta_{r+N} &= \sum_{j=1}^{N} \int_{\mu_j^o}^{\mu_j} \left(\frac{1}{(\mu_j - \bar{a}_k)} - \frac{1}{(\mu_j - \bar{a}_r)} \right) d\mu_j \\
&\quad + \sum_{j=1,j\neq k}^{N} \int_{\mu_{j+N}^o}^{\mu_{j+N}} \left(\frac{1}{(\mu_{j+N} - \bar{a}_k)} - \frac{1}{(\mu_{j+N} - \bar{a}_r)} \right) d\mu_{j+N} \\
&= 2i(\bar{a}_k - \bar{a}_r)(x + v_{r+N}t),
\end{aligned}
\tag{3.36}
$$

and for $N \geq r > k$, let

$$
\begin{aligned}
\theta_{r+N} &= \sum_{j=1}^{N} \int_{\mu_j^o}^{\mu_j} \left(\frac{1}{(\mu_j - a_k)} - \frac{1}{(\mu_j - a_r)} \right) d\mu_j \\
&\quad - \sum_{j=1,j\neq k}^{N} \int_{\mu_{j+N}^o}^{\mu_{j+N}} \left(\frac{1}{(\mu_{j+N} - a_k)} - \frac{1}{(\mu_{j+N} - a_r)} \right) d\mu_{j+N} \\
&= 2i(a_k - a_r)(x + v_{r+N}t).
\end{aligned}
\tag{3.37}
$$

Here

$$
\left.
\begin{aligned}
v_{r+N} &= \bar{a}_k + \bar{a}_r \quad \text{for } r < k, \\[2mm]
v_{r+N} &= a_k + a_r \quad \text{for } r > k.
\end{aligned}
\right\}
\tag{3.38}
$$

Now we take the real parts of the expressions (3.30), (3.33), (3.34), (3.36) and (3.37) and apply asymptotic reduction along the direction d_k ($1 \leq k \leq N$). This yields the following limits for the remaining $(N-1)$ μ_r-variables:

$$
\left.
\begin{aligned}
&1. \quad t \to \infty \; : \; \mu_{r+N} \to \bar{a}_r, \; \text{for } r < k \\[2mm]
&\quad \text{and } \mu_{r+N} \to a_r \text{ for } N \geq r > k \\[2mm]
&2. \quad t \to -\infty \; : \; \mu_{r+N} \to a_r \text{ for } r < k \\[2mm]
&\quad \text{and } \mu_{r+N} \to \bar{a}_r \text{ for } r > k.
\end{aligned}
\right\}
\tag{3.39}
$$

After substituting these values into the expression for the k-th angle variable it is transformed into the 1-soliton angle variable with some acquired complex phase:

$$
\left.
\begin{aligned}
\Theta_k^+ &= \frac{1}{4\mathrm{Im}(a_k)} \int_{\mu_k^o}^{\mu_k} \left(\frac{1}{(\mu_k - a_k)} - \frac{1}{(\mu_k - \bar{a}_k)} \right) d\mu_k + \varphi_k^+ \\[2mm]
&= x + v_k t, \\[4mm]
\Theta_k^- &= -\frac{1}{4\mathrm{Im}(a_k)} \int_{\mu_k^o}^{\mu_k} \left(\frac{1}{(\mu_k - a_k)} - \frac{1}{(\mu_k - \bar{a}_k)} \right) d\mu_k + \varphi_k^- \\[2mm]
&= x + v_k t.
\end{aligned}
\right\}
\tag{3.40}
$$

Here, basic points of the integrals are taken on different sheets of the Riemann surface. $\square$

Now using the phase function and the general definition of a geometric phase given above we calculate a real part of the complex (f)NLS phase.

Corollary 3.4 *Consider the difference between Θ_k^- and Θ_k^+ and let $d_k \to \infty$. This yields a phase generated by the singularities of the initial system of equations:*

$$\Delta_G\Theta_k = -\frac{1}{2\mathrm{Im}(a_k)}\oint_{L_k}\left(\frac{1}{(\mu_k-a_k)}-\frac{1}{(\mu_k-\bar{a}_k)}\right)d\mu_k = \varphi_k^- - \varphi_k^+$$

$$= \frac{1}{2\mathrm{Im}(a_k)}\sum_{j=1}^{k-1}\log\left|\frac{a_j-\bar{a}_k}{a_j-a_k}\right| - \frac{1}{2\mathrm{Im}(a_k)}\sum_{j=k+1}^{N}\log\left|\frac{a_j-\bar{a}_k}{a_j-a_k}\right|. \quad (3.41)$$

Here, the integral is taken along the cycle L_k over the basic cut on the Riemann surface.

The method of asymptotic reduction of the angle representations described above yields the splitting of (3.30) into (N) 1-soliton angle representations with their corresponding phase functions. Therefore, the angle representation (3.30) describes N-soliton solutions.

Corollary 3.5 *The first part (3.30) of the angle representation is the same for all directions d_k. The choice of the last $(N-1)$ elements of the representation as well as the choice of the ϵ_{rj} is an important part of the asymptotic reduction in case of the (f)NLS equation. The problem is considered on the linear space of the angle representations. Namely, along with every fixed direction d_k, we choose a new particular set of angle variables $(\theta_{r+N}, \ r=1,...,N, \ r \neq k)$.*

To deal with the limiting imaginary part of (3.30), we use the formulae connecting the solution Q of the (f)NLS to the function U:

$$U = i\frac{\partial}{\partial x}\log Q, \quad U = -2\sum_{j=1}^{g}\mu_j + 2\sum_{k=1}^{N}(a_k + \bar{a}_k). \quad (3.42)$$

Integrating this expression, we obtain an arbitrary constant in the phase of the exponent. We put this constant equal to the argument part of the phase corresponding to (3.41). (See the 1-dimensional case discussed in Section 3.1.)

The angle representations introduced above demonstrate the existence of the residual interaction between solitons after the asymptotic splitting of N-soliton solutions of the (f)NLS. This residual interaction is a manifestation of the degeneracy of the invariant varieties in the phase space of this soliton problem. It generates the ghost solitons described in previous section.

4 Asymptotic reduction and sine-Gordon geometric phases

Another important problem that can be investigated by the method of complex angle representations is the description of the interaction between different types of soliton solutions of the same nonlinear equation. To demonstrate this, we will investigate the breather-kink-antikink interaction and the effect of this interaction on the separatrix of the sine-Gordon equation and corresponding geometric phases. In Bishop, McLaughlin and Solerno [1989], 2-dimensional real angle variables were introduced to describe a transformation of a breather into a kink-antikink solution through a separatrix. In the present paper, we are investigating an interaction between different types of soliton solutions of the sine-Gordon equation and their effect on the transition through an n-dimensional separatrix.

We consider $(N = (n_1 + 2n_2))$-dimensional complex phase space $\mathbf{C}^N$ and investigate general soliton solutions using asymptotic reduction of the combination of the n_1-dimensional breather and n_2-dimensional kink-antikink angle representations. The general approach yields an interesting new class of phases and describes the transition through separatrix solution in the n-dimensional case.

To clarify the link with algebraic geometry we begin with the quasiperiodic solutions. (We will be using notations introduced in the previous section for (f)NLS equation.) In the case of the sine-Gordon equation, the associated quasiperiodic system has the form

$$\frac{\partial \mu_j}{\partial x} = 2(1 + G_{N-1}(\mu_j))\frac{\sqrt{-\mu_j C(\mu_j)}}{\prod_{i\neq j}(\mu_j - \mu_i)}, \tag{4.1}$$

$$\frac{\partial \mu_j}{\partial t} = 2(1 - G_{N-1}(\mu_j))\frac{\sqrt{-\mu_j C(\mu_j)}}{\prod_{i\neq j}(\mu_j - \mu_i)}, \quad j = 1, ..., N, \tag{4.2}$$

defined on the symmetric product of the n copies of the hyperelliptic curve

$$W^2 = -\frac{C(\mu)}{\mu}, \quad \text{where } C(\mu) = \prod_{r=1}^{2N}(\mu - m_r) \tag{4.3}$$

is a spectral polynomial. (See M. Alber and S. Alber [1985] and Ercolani, Forest, McLaughlin, and Montgomery [1987].) Here

$$G_{N-1}(\mu) = \frac{1}{4\sqrt{\prod_{r=1}^{2N} m_r}} \sum_{l=0}^{N-1} g_l \mu^{N-l-1} \tag{4.4}$$

with the coefficients from

$$G(\mu) = \prod_{j=1}^{N}(\mu - \mu_j) = \sum_{r=0}^{N} g_r \mu^{N-r}. \tag{4.5}$$

Using the different limiting processes:

$$m_{2r}, m_{2r-1} \to a_r, \quad \text{where } a_r = -\alpha_r^2, \quad r = 1, ..., n_1, \tag{4.6}$$

which gives an (n_1) kink-antikink solution

and

$$m_{n_1+4r}, m_{n_1+4r-1} \to h_r, \ m_{n_1+4r-2}, m_{n_1+4r-3} \to \bar{h}_r, \quad r = 1, ..., n_2, \tag{4.7}$$

which gives an (n_2) breather solution

applied to different parts of the discrete spectrum (m_r) one can obtain a combination of soliton solutions: breathers and kinks. It leads to the transformation of the basic polynomial: $\sqrt{C(\mu)} \to \bar{C}(\mu)$, which results in the transformation of the differential systems (4.1) and (4.2).

4.1 Breather-kink collision and corresponding phase shifts The idea is to use different pieces of the soliton and resonant angle representations to construct representation for the collision in the form of a mosaic defined on the Riemann surface of rather complicated structure. Different pieces are kept together through the same set of μ-variables. Note that the general form of each piece corresponding to a particular type of soliton solution is preserved. What is changed is the dimension of the underlying Riemann surface which is determined by the dimension of the mosaic.

We illustrate the action angle variables produced by our method for the two soliton case of the breather-kink-antikink collision, which corresponds to the following discrete spectrum: $(a_1, a_2; h, \bar{h})$. Here $a_1, a_2 \in \mathbf{R}$ and $h, \bar{h} \in \mathbf{C}$.

$$I_1 = a_1, \quad \theta_1 = \frac{1}{2} \sum_{j=1}^{4} \int_{\mu_j^o}^{\mu_j} \frac{1}{(\mu_j - a_1)} \frac{d\mu_j}{\sqrt{-\mu_j}} = w_1 x + v_1 t, \qquad (4.8)$$

$$I_2 = a_2, \quad \theta_2 = \frac{1}{2} \sum_{j=1}^{4} \int_{\mu_j^o}^{\mu_j} \frac{1}{(\mu_j - a_2)} \frac{d\mu_j}{\sqrt{-\mu_j}} = w_2 x + v_2 t, \qquad (4.9)$$

and

$$I_3 = h, \quad \theta_3 = \frac{1}{2} \sum_{j=1}^{4} \int_{\mu_j^o}^{\mu_j} (\frac{1}{(\mu_j - h)} \frac{d\mu_j}{\sqrt{-\mu_j}} = w_3 x + v_3 t, \qquad (4.10)$$

$$I_4 = \bar{h}, \quad \theta_4 = \frac{1}{2} \sum_{j=1}^{4} \int_{\mu_j^o}^{\mu_j} \frac{1}{(\mu_j - \bar{h})} \frac{d\mu_j}{\sqrt{-\mu_j}} = \bar{w}_3 x + \bar{v}_3 t. \qquad (4.11)$$

Here

$$w_r = \left(1 - \frac{1}{4a_r}\right), \quad v_r = \left(1 + \frac{1}{4a_r}\right) \quad r = 1, 2 \qquad (4.12)$$

and

$$w_3 = \left(1 - \frac{1}{4h}\right), \quad v_3 = \left(1 + \frac{1}{4h}\right). \qquad (4.13)$$

4.2 n-dimensional case *The systems obtained from (4.1) and (4.2) as a result of the limiting procedures (4.6) and (4.7) can be checked to be Hamiltonian systems with the Hamiltonians*

$$H_s^s = \frac{\sum_{j=1}^{n}(e^{2\sqrt{-\mu_j}P_j} - \bar{C}(\mu_j))(1 + G_{N-1}(\mu_j))}{\prod_{l \neq j}(\mu_j - \mu_l)} \qquad (4.14)$$

and

$$H_s^d = \frac{\sum_{j=1}^{n}(e^{2\sqrt{-\mu_j}P_j} - \bar{C}(\mu_j))(1 - G_{N-1}(\mu_j))}{\prod_{l \neq j}(\mu_j - \mu_l)}. \qquad (4.15)$$

These systems have a complete set of first integrals

$$P_j = \frac{\log \bar{C}(\mu_j)}{2\sqrt{-\mu_j}}, \quad j = 1, ..., N \tag{4.16}$$

and a system of action-angle variables. Here $N = n_1 + 2n_2$.

The action-angle variables can be described as follows:

$$I_r = a_r, \quad \theta_r = -\frac{\partial S}{\partial I_r} = \frac{1}{2} \sum_{j=1}^{N} \int_{\mu_j^o}^{\mu_j} \frac{1}{(\mu_j - a_r)} \frac{d\mu_j}{\sqrt{-\mu_j}} = w_r x + v_r t, \quad r = 1, ..., n_1, \tag{4.17}$$

and

$$I_{r+n_1} = h_r, \quad \theta_{r+n_1} = \frac{1}{2} \sum_{j=1}^{N} \int_{\mu_j^o}^{\mu_j} (\frac{1}{(\mu_j - h_r)} \frac{d\mu_j}{\sqrt{-\mu_j}} = w_r x + v_r t, \quad r = 1, .., n_2, \tag{4.18}$$

$$I_{r+n_2+n_1} = \bar{h}_r, \quad \theta_{r+n_2+n_1} = \frac{1}{2} \sum_{j=1}^{N} \int_{\mu_j^o}^{\mu_j} \frac{1}{(\mu_j - \bar{h}_r)} \frac{d\mu_j}{\sqrt{-\mu_j}} = \bar{w}_r x + \bar{v}_r t, \tag{4.19}$$

$$r = 1, .., n_2.$$

Here

$$w_r = \left(1 - \frac{1}{4a_r}\right), \quad v_r = \left(1 + \frac{1}{4a_r}\right) \quad r = 1, ..., n_1 \tag{4.20}$$

or

$$w_r = \left(1 - \frac{1}{4h_r}\right), \quad v_r = \left(1 + \frac{1}{4h_r}\right) \quad r = 1, ..., n_2. \tag{4.21}$$

Note that the soliton (kink-antikink) angle representation for the sine-Gordon equation is similar to a particular case ($n_1 = 2n$) of the KdV and (d)NLS representation investigated in Alber and Marsden [1992]. Therefore, it remains only to describe in detail the pure breather angle representation.

We change variables as follows

$$\mu_j = -\xi_j^2, \quad h_r = -\chi_r^2, \tag{4.22}$$

choose ϵ_{rj} in a way specified below to find that real part of the angle representation is represented in the form

$$\theta_r = \frac{2|h_r|^2}{\text{Re}(\chi_r)(4|h_r|^2 - \text{Re}(h_r))} \sum_{j=1}^{n_2} \left(\log \left| \frac{\xi_j - \chi_r}{\xi_j + \chi_r} \right| \right) \Big|_{\xi_j^0}^{\xi_j}$$

$$- \frac{2|h_r|^2}{\text{Re}(\chi_r)(4|h_r|^2 - \text{Re}(h_r))} \sum_{j=n_2+1}^{2n_2} \left(\log \left| \frac{\xi_j - \chi_r}{\xi_j + \chi_r} \right| \right) \Big|_{\xi_j^0}^{\xi_j}$$

$$= x + \frac{(4|h_r|^2 + \text{Re}(h_r))}{(4|h_r|^2 - \text{Re}(h_r))} t, \quad r = 1, ..., n_2 \tag{4.23}$$

and

$$\theta_{r+n_2} = \frac{2|h_r|^2}{\operatorname{Re}(\chi_r)(4|h_r|^2 - \operatorname{Re}(h_r))} \sum_{j=1}^{2n_2} \left(\log \left| \frac{\xi_j - \bar{\chi}_r}{\xi_j + \bar{\chi}_r} \right| \right) \Big|_{\xi_j^0}^{\xi_j}$$

$$= x + \left(\frac{4|h_r|^2 + \operatorname{Re}(h_r)}{4|h_r|^2 - \operatorname{Re}(h_r)} \right) t. \tag{4.24}$$

Now we apply asymptotic reduction to the real parts of the first n_2 elements of the angle representation. This yields the following distribution of the limiting values of the μ-variables along real directions d_k:

$$d_k = x + V_k t = x + \left(\frac{4|h_k|^2 + \operatorname{Re}(h_k)}{4|h_k|^2 - \operatorname{Re}(h_k)} \right) t$$

$$\left. \begin{aligned} &1. \quad t \to \infty \; : \; \mu_r \to \chi_r, \;\; \text{for } V_r < V_k \;\; \text{and} \;\; \mu_r \to \bar{\chi}_r \text{ for } V_r > V_k \\[2mm] &2. \quad t \to -\infty \; : \; \mu_r \to \bar{\chi}_r \text{ for } V_r < V_k \;\; \text{and} \;\; \mu_r \to \chi_r \text{ for } V_r > V_k. \end{aligned} \right\} \tag{4.25}$$

Here it is useful to keep in mind that $\mu_{j+n_2} = \bar{\mu}_j$ and $V_{n_2} < ... < V_1$.

Now we combine angle representations and obtain phases acquired by breathers as a result of interaction with kinks and vice versa. This results in the following phase shift:

$$\begin{aligned} \Delta_G \Theta_k &= \frac{2|h_k|^2}{\operatorname{Re}(h_k)(4|h_k|^2 - \operatorname{Re}(h_k))} \sum_{j=1}^{k-1} \log \left| \frac{(\chi_j - \chi_k)(\chi_j + \bar{\chi}_k)}{(\chi_j + \chi_k)(\chi_j - \bar{\chi}_k)} \right| \\[3mm] &\quad - \frac{2|\chi_k|^2}{\operatorname{Re}(h_k)(4|h_k|^2 - \operatorname{Re}(h_k))} \sum_{j=k+1}^{n_2} \log \left| \frac{(\chi_j - \chi_k)(\chi_j + \bar{\chi}_k)}{(\chi_j + \chi_k)(\chi_j - \bar{\chi}_k)} \right| \\[3mm] &\quad + \Delta_G \Theta_k^{\text{breather-kink}}, \end{aligned}$$

where $\Delta_G \Theta_k^{\text{breather-kink}}$ is the phase acquired by the breather after interaction with the kink solution and is given by the following expression:

$$\begin{aligned} \Delta_G \Theta_k^{\text{breather-kink}} &= \frac{2|h_k|^2}{\operatorname{Re}(h_k)(4|h_k|^2 - \operatorname{Re}(h_k))} \sum_{j=1,(v_j/w_j)<V_k}^{n_2} \log \left| \frac{(\alpha_j - \chi_k)(\alpha_j + \bar{\chi}_k)}{(\alpha_j + \chi_k)(\alpha_j - \bar{\chi}_k)} \right| \\[3mm] &\quad - \frac{2|\chi_k|^2}{\operatorname{Re}(h_k)(4|h_k|^2 - \operatorname{Re}(h_k))} \sum_{j=1,(v_j/w_j)>V_k}^{n_2} \log \left| \frac{(\alpha_j - \chi_k)(\alpha_j + \bar{\chi}_k)}{(\alpha_j + \chi_k)(\alpha_j - \bar{\chi}_k)} \right|, \end{aligned}$$

where $a_j = -\alpha_j^2$. These phases depend on the mutual distribution of the elements of the breather and kink parts of the discrete spectrum.

The additional phase that is acquired by the kink along

$$d_k = x + \frac{v_k}{w_k}$$

due to interaction with breathers has the form:

$$
\Delta_G \Theta_k^{\text{kink-breather}} \;=\; \frac{2|\alpha_k|^2}{\operatorname{Re}(\alpha_k)(4|a_k|^2 - 1))} \sum_{j=1,\,V_j<(v_k/w_k)}^{n_1} \log\left|\frac{(\chi_j - \alpha_k)}{(\chi_j + \alpha_k)}\right|
$$

$$
- \; \frac{2|\alpha_k|^2}{\operatorname{Re}(\alpha_k)(4|a_k|^2 - 1))} \sum_{j=1,\,V_j>(v_k/w_k)}^{n_1} \log\left|\frac{(\chi_j - \alpha_k)}{(\chi_j + \alpha_k)}\right|.
$$

Phases for the imaginary parts of the angle representation are obtained in the form of argument shifts corresponding to the same logarithmic phase function.

In the case of complex geometric phases, the real and imaginary parts of the phase usually have different physical meanings. For example, in the case of a breather solution of the sine-Gordon equation, these two parts are related to two different types of oscillations of the breather; one is a wave phase oscillation and the other is an amplitude oscillation. For the (f)NLS equation, they correspond to temporal and spatial phase shifts of optical solitons.

Above, we considered asymptotic reduction along real "breather" and "kink" directions. Note that complex angle representations obtained in this paper can be also used for integrating complex Hamiltonian flows along complex directions in x and t.

5   Resonant geometric phases

Below we introduce $2n$-dimensional angle representations for resonant solutions of the sine-Gordon equation and describe resonant geometric phases. This technique is of course not restricted to the sine-Gordon equation, but can be applied to any soliton equation. Resonant solutions are considered as singular points in the moduli space of solitons and correspond to multiple points of the discrete spectrum.

As we remarked in the introduction, as one passes from n soliton solutions to resonant solitons, the spectrum changes continuously, but the invariant variety on which the solitons reside, does not. Thus, one has to use the tools of any perturbative set up with great care. The methods of asymptotic reduction provide these tools.

We will show that asymptotic reduction of the resonant angle representations along a 2-dimensional direction splits them into the sum of n 2-dimensional resonant solutions. This kind of asymptotic splitting of solitons was previously known only in the case of n-soliton solutions splitting into single solitons, and only along a one dimensional direction. The main difference is that in the present case, the 2-dimensional resonant angle representations are irreducible in the sense that they do not split asymptotically.

In particular, this approach yields a detailed description of the complex phases for the (f)NLS and sine-Gordon equations, including phases for the breather-kink and soliton-separatrix interactions.

A resonant solution can be considered as a separatrix between different soliton solutions of the same nonlinear equation. We will demonstrate this idea using the breather-kink-antikink interaction and the transition through a separatrix in the case of the sine-Gordon equation.

Soliton kink-antikink and breather solutions of the sine-Gordon equation are separated by the separatrix solutions obtained as follows. Namely, one takes in

(4.17) $n_1 = 2n_3$ and considers the following limiting process:

$$(a_{2k}, a_{2k-1}) \to b_k, \quad k = 1, ..., n_3.$$

This provides an example of a resonant solution.

The resonant solution (i.e., separatrix solution) in the case of the sine-Gordon equation can be described by the angle representation

$$\left.\begin{aligned}
\theta_r &= \frac{1}{2} \sum_{j=1}^{2n_3} \int_{\mu_j^o}^{\mu_j} \frac{1}{(\mu_j - b_r)} \frac{d\mu_j}{\sqrt{-\mu_j}} = w_r x + v_r t + \theta_r^0, \\[2em]
\theta_{r+n_3} &= \frac{1}{2} \sum_{j=1}^{2n_3} \int_{\mu_j^o}^{\mu_j} \frac{1}{(\mu_j - b_r)^2} \frac{d\mu_j}{\sqrt{-\mu_j}} = \rho_r x + \nu_r t + \theta_{r+n_3}^0, \\[2em]
r &= 1, ..., n_3.
\end{aligned}\right\} \tag{5.1}$$

Here

$$w_r = \left(1 + \frac{1}{b_r}\right), \ v_r = \left(1 - \frac{1}{b_r}\right), \ \rho_r = -\frac{1}{b_r^2}, \ \nu_r = \frac{1}{b_r^2}.$$

The first n_3 integrals can be evaluated as before using change of variables

$$\mu_j = -\xi_j^2, \ b_r = -\beta_r^2 \tag{5.2}$$

resulting in the introduction of the logarithmic phase functions. The phase function for the second part of (5.1) can be obtained using the fact that

$$\begin{aligned}
\theta_{r+n_3} &= -\frac{\partial \theta_r}{\partial b_r} = \frac{1}{2\beta_r} \frac{\partial \theta_r}{\partial \beta_r} \\[2em]
&= \frac{1}{2(2\beta_r)^2} \sum_{j=1}^{2n_3} \left(\frac{1}{|\xi_j - \beta_r|} - \frac{1}{|\xi_j + \beta_r|}\right)\Bigg|_{\xi_j^0}^{\xi_j} - \frac{1}{8(\beta_r)^2}\theta_r \\[2em]
&= \theta_{r+n_3}^0 + \frac{1}{b_r^2}(t - x).
\end{aligned} \tag{5.3}$$

We apply asymptotic reduction by investigating the following system, obtained from the resonant angle representation described above,

$$\theta_r = \frac{1}{4\beta_r(1 - b_r)} \sum_{j=1}^{2n_3} \epsilon_{rj} \log \left|\frac{\xi_j - \beta_r}{\xi_j + \beta_r}\right| \Bigg|_{\xi_j^0}^{\xi_j} = x + V_r t, \quad V_r = \frac{b_r - 1}{b_r + 1}, \tag{5.4}$$

$$\theta_{r+n_3} = \frac{1}{8\beta_r^3} \sum_{j=1}^{2n_3} \epsilon_{rj} \left(\frac{1}{|\xi_j - \beta_r|} - \frac{1}{|\xi_j + \beta_r|}\right)\Bigg|_{\xi_j^0}^{\xi_j} = \frac{1}{b_r^2}(x - t) + \frac{1}{8\beta_r^2}(x + V_r t),$$

$$\tag{5.5}$$

$$r = 1, ..., n_3.$$

Remark 5.1 Note that we are dealing with $(2n_3)$ μ_j variables and only (n_3) action variables b_j. This means that we consider each pair of variables μ_j and μ_{j+n_3} on the two copies of the same Riemann surface with the branch point b_j.

The equations corresponding to θ_r and θ_{r+n_3} from the systems (5.4) and (5.5) determine exponential and rational rates of convergence to $\pm\beta_r$ for ξ_r and ξ_{r+n_3} along directions d_r and d_{r+n_3} respectively

$$d_r = x + V_r t, \quad d_{r+n_3} = \frac{1}{b_r^2}(x - t) + \frac{1}{8\beta_r^2}(x + V_r t).$$

Now we fix two directions d_k and d_{k+n_3} and investigate the asymptotic limit of every ξ_j for $t \to \pm\infty$.

As a result of the asymptotic reduction, we obtain a 2-dimensional separatrix solution that consists of one logarithmic and one rational type of angle variable, together with the acquired phase function φ. The logarithmic angle variable is

$$
\begin{aligned}
\Theta_k^{+,-} &= \frac{b_k}{2(1+b_k)} \int_{\mu_k^o}^{\mu_k} \frac{1}{(\mu_k - b_k)} \frac{d\mu_k}{\sqrt{-\mu_k}} \\
&+ \frac{b_k}{2(1+b_k)} \int_{\mu_{k+n_3}^o}^{\mu_{k+n_3}} \frac{1}{(\mu_{k+n_3} - b_k)} \frac{d\mu_{k+n_3}}{\sqrt{-\mu_{k+n_3}}} - \varphi_k^{+,-} = x + V_k t,
\end{aligned}
$$

$$(5.6)$$

and the rational one is

$$
\begin{aligned}
\Theta_{k+n_3}^{+,-} &= \frac{1}{2} \int_{\mu_k^o}^{\mu_k} \frac{1}{(\mu_k - b_k)^2} \frac{d\mu_k}{\sqrt{-\mu_k}} \\
&+ \frac{1}{2} \int_{\mu_{k+n_3}^o}^{\mu_{k+n_3}} \frac{1}{(\mu_{k+n_3} - b_k)^2} \frac{d\mu_{k+n_3}}{\sqrt{-\mu_{k+n_3}}} - \varphi_{k+n_3}^{+,-} = \rho_k x + \nu_k t.
\end{aligned}
$$

This also results in the accumulation of the following resonant phases:

$$
\begin{aligned}
\Delta_G \Theta_k &= \varphi_k^+ - \varphi_k^- = \frac{1}{2\beta_k(1-b_k)} \sum_{j=1, V_j < V_k}^{n_3} \log\left|\frac{\beta_j - \beta_k}{\beta_j + \beta_k}\right| \\
&- \frac{1}{2\beta_k(1-b_k)} \sum_{j=1, V_j > V_k}^{n_3} \log\left|\frac{\beta_j - \beta_k}{\beta_j + \beta_k}\right|,
\end{aligned}
$$

$$
\begin{aligned}
\Delta_G \Theta_{k+n_3} &= \varphi_{k+n_3}^+ - \varphi_{k+n_3}^- = \frac{1}{8\beta_k^3} \sum_{j=1, V_j < V_k}^{n_3} \left(\frac{1}{|\beta_j - \beta_k|} - \frac{1}{|\beta_j + \beta_k|}\right) \\
&- \frac{1}{8\beta_k^3} \sum_{j=1, V_j > V_k}^{n_3} \left(\frac{1}{|\beta_j - \beta_k|} - \frac{1}{|\beta_j + \beta_k|}\right).
\end{aligned}
$$

Remark 5.2 Note that the angle representation can be smoothly transformed from breather case to kink-antikink case. The problem is that the system of action variables is degenerate for the resonant (separatrix) solution considered on the same phase space $\mathbf{C}^N$ together with breathers and kinks. Here $N = n_1 + 2n_2 + 2n_3$.

On the other hand, we have enough angle variables to describe the resonant (separatrix) subspace of the phase space and to calculate corresponding phases.

Now we consider a combination of all elements of the angle representation and link them to each other through the same set of exactly $N = n_1 + 2n_2 + 2n_3$ variables μ_j. Here, each of the μ_j is on the corresponding Riemann surface.

Remark 5.3 Asymptotic reduction developed in this paper for resonant solutions of nonlinear equations enables one to construct and investigate n-dimensional umbilic solitons introduced in Alber and Marsden [1994]. It also yields angle representations for the solitons with quasiperiodic background and for m-dimensional complex solutions which describe collision of m solitons moving with the same speed.

6 Discrete asymptotic reduction and geometric phases

Finally, we give an example of the application of asymptotic reduction to the Toda lattice regarded as a spatial discretization of the defocussing (d)NLS equation. We will do this on the level of the angle representations for the two problems. To make this link, we will choose directions for the asymtotic reduction that are partially discrete. In fact, we will choose directions appropriate to solitons for (d)NLS, and along these directions, we leave the time variable t continuous, but choose the spatial variable x to be discrete.

In particular, our approach provides angle representations and exponential Hamiltonians for the N soliton solutions of the infinite particle Toda lattice (on the line) and gives a clear understanding of the corresponding geometric phases. We will also show that asymptotic reduction of the N-soliton angle representation for the Toda lattice yields a splitting into the sum of 1-dimensional solitons similar to the solitons of "finite density" of the (d)NLS. (For details about a connection between Toda lattices and continuous problems, see Toda [1989] and S. Alber [1989, 1991].)

Theorem 6.1 *N-soliton solutions for the infinite particle Toda lattice system are described by the following angle representation*

$$\theta_k = \sum_{j=1}^{N} \int_{\mu_j^0}^{\mu_j} \frac{d\mu_{j,k}}{(a_k - \mu_{j,k})\sqrt{(\mu_{j,k} - b_1)(\mu_{j,k} - b_2)}} = t + rh_k + \theta_k^0, \tag{6.1}$$

$$k = 1, ..., N; \quad N = 2n + 1,$$

where (r) is a discrete variable. The system (6.1) splits into the sum of 1-dimensional angle representations of the same type as $t \to \pm\infty$.

Proof One first works out the Hamiltonian system and associated angle representation for the quasiperiodic case as one does for the NLS equation, as was described before, or in Alber and Marsden [1992]. Then one passes to the N soliton limit by collapsing the spectrum as before. This leads to a new system of angle representations as stated in the theorem. To study the asymptotic splitting, we prepare the following lemma.

Lemma 6.2 *The 1-dimensional angle representation*

$$\theta = \int_{\mu^0}^{\mu} \frac{d\mu}{(a - \mu)\sqrt{(\mu - b_1)(\mu_k - b_2)}} = t + rh + \theta^0 \tag{6.2}$$

corresponds to a class of solutions of the infinite particle Toda lattice, which includes the 1-soliton solutions.

Proof We introduce new variables

$$\xi^2 = \frac{\mu - b_1}{\mu - b_2}, \quad \mu = \frac{\xi^2 b_2 - b_1}{\xi^2 - 1}. \tag{6.3}$$

This results in the following form of θ:

$$\theta = \frac{1}{\rho} \log \left| \frac{\xi - \alpha}{\xi + \alpha} \right| = t + rh + \theta^0, \quad r \in Z, \tag{6.4}$$

where

$$\alpha^2 = -\frac{b_1 - a}{a - b_2}, \quad \rho = \sqrt{(a - b_1)(a - b_2)}. \tag{6.5}$$

From this it follows that

$$\mu = b_2 + \frac{(b_2 - b_1)}{2} \left(\frac{(1 - \beta)}{\alpha(1 + \beta) - (1 - \beta)} - \frac{(1 - \beta)}{\alpha(1 + \beta) + (1 - \beta)} \right), \tag{6.6}$$

where

$$\beta = \exp(\rho\theta^0 + \rho t + \rho r h) = \exp(\rho\zeta). \tag{6.7}$$

This yields the solution (q) of the Toda lattice in the form

$$q = C_0 + \log \left(\frac{1 + e^\nu e^{\rho\zeta}}{1 + e^{-\nu} e^{\rho\zeta}} \right), \quad \nu = \log \left(\frac{\alpha + 1}{\alpha - 1} \right), \tag{6.8}$$

which is the Toda 1-soliton solution. The proof of the splitting is similar to the continuous case. The main difference between the discrete and continuous cases is the description of a "fixed direction" D_k. Note that we are considering Toda flow (which is linearized in terms of the angle representation) as an "exact" spatial discretization of the soliton flow of the (d)NLS equation on the noncompact invariant variety described in Alber and Marsden [1992]. Therefore, we consider asymptotic reduction for the infinite particle Toda lattice along a spatial discretization of the continuous direction d_k used before for the (d)NLS equation. In particular, this yields discrete geometric phases. $\square$

Corollary 6.3 *The 1-dimensional angle representation in the above lemma corresponding to the case when $b_1 = b_2 = b$, namely,*

$$\theta = \int_{\mu^0}^{\mu} \frac{d\mu}{(a - \mu)(\mu - b)} = t + rh + \theta^0 = \frac{\beta}{a - b} \tag{6.9}$$

yields an expression for q, namely

$$q = bt + \log \left(1 + e^\beta \right) + C_0, \tag{6.10}$$

which is similar to a "dark" soliton of the (d)NLS equation.

Angle representations of this type belong to the same class as action-angle variables first introduced in Flaschka and McLaughlin [1976].

In a forthcoming paper we will describe new types of discretizations of the soliton problem for the (f)NLS equation which can be investigated using the method of this section. Soliton solutions of these new discrete systems can be considered as an approximation (similar to infinite particle Toda lattice models) for optical solitons.

Lastly, angle representations obtained in Alber and Marsden [1994] for homoclinic orbits of the C. Neumann problem (extended to the discrete case by the method of this section) can be used to describe corresponding homoclinic orbits of the stationary Heisenberg chain with classical spins. (For details concerning a connection between discrete version of the C. Neumann problem and the Heisenberg chain see Moser and Veselov [1991]).

Acknowledgements. Mark Alber thanks The Fields Institute for its kind hospitality during two visits in 1993. We also thank Dave McLaughlin for several helpful suggestions.

References

Ablowitz, M.J., and Segur, H. [1981] *Solitons and the Inverse Scattering Transform*, SIAM, Philadelphia.

Adler, M., and Van Moerbeke, P. [1980] *Completely integrable systems Kac-Moody Lie algebras and curves*, Adv. in Math., **38**, 267–317.

Alber, M.S. [1992] *Complex Geometric Asymptotics, Geometric Phases and Nonlinear Integrable Systems*; in Studies in Math. Phys. 3, North-Holland, Elsevier Science Publishers B.V., Amsterdam, 415–427.

――――[1991] *Hyperbolic Geometric Asymptotics*, Asymptotic Analysis, **5**, 161–172.

Alber, M.S., and Alber, S.J. [1985] *Hamiltonian formalism for finite-zone solutions of integrable equations*, C.R. Acad. Sc. Paris, **301**, 777–781.

――――[1987] *Hamiltonian formalism for nonlinear Schrödinger equations and sine-Gordon equations*, J. London Math. Soc., **36**, 176–192.

Alber, M.S., and Marsden, J.E. [1992] *On Geometric Phases for Soliton Equations*, Commun. Math. Phys., **149**, 217–240.

――――[1994] *Geometric Phases and Monodromy at Singularities*, NATO Advanced Study Institute, Series C, (to appear).

Alber, M.S., Camassa, R., Holm, D. and Mardsen, J.E. [1994a] *The geometry of peaked solitons and billiard solutions of a class of integrable pde's*, Lett. Math. Phys. (to appear).

――――[1994b] *On the geometry of umbilic solitons* (to appear).

Alber, S.J. [1989] *On finite-zone solutions of the relativistic Toda lattices*, Lett. Math. Phys., **17**, 149–155.

――――[1991] *Associated integrable systems*, J. Math. Phys., **32**, 916–922.

Berry, M.V. [1984] *Quantal phase factors accompanying adiabatic changes*, Proc. R. Soc. Lond. A, **392**, 45–57.

Berry, M.V., and Hannay, J.H. [1988] *Classical non-adiabatic angles*, J. Phys. A: Math. Gen., **21**, L325–L331.

Birnir, B. [1986] *Singularities of the complex Korteweg-de Vries flows*, Comm. Pure Appl. Math., **39**, 283–305.

Bishop, D.W., McLaughlin, D., and Solerno, M. [1989] *Global coordinates for the breather-kink (antikink) sine-Gordon phase space: An explicit separatrix as a possible source of chaos*, Phys. Rev. A, **40**, 6463–6469.

Camassa, R., and Holm, D.D. [1993] *An integrable shallow water equation with peaked solitons*, Phys. Rev. Lett., **71**, 1661–1664.

Camassa, R., Holm, D.D., and Hyman, J.M. [1993] *A new integrable shallow water equation*, Adv. in Appl. Mech., (to appear).

David, D., and Holm, D.D. [1992] *Multiple Lie-Poisson Structures, Reductions, and Geometric Phases for the Maxwell-Bloch Traveling Wave Equation*, J. Nonlinear Sci., **2**, 241–262.

Devaney, R. [1978] *Transversal homoclinic orbits in an integrable system*, Amer. J. Math., **100**, 631.

Duistermaat, J.J. [1980] *On global Action-Angle Coordinates*, Comm. Pure Appl. Math., **23**, 687–706.

Ercolani, N. [1989] *Generalized Theta functions and homoclinic varieties*, Proc. Symp. Pure Appl. Math., **49**, 87–100.

Ercolani, N., Forest, M., McLaughlin, D.W., and Montgomery, R. [1987] *Hamiltonian structure for the modulation equations of a sine-Gordon wavetrain*, Duke Math. Journal, **55**, 949–983.

Ercolani, N., and Forest, M. [1985] *The Geometry of Real Sine-Gordon Wavetrains*, Commun. Math. Phys., **99**, 1–49.

Ercolani, N., and McLaughlin, D.W. [1991] *Toward a topological classification of integrable PDE's*; in Proc. of a Workshop on The Geometry of Hamiltonian Systems (T. Ratiu, ed.), MSRI Publications, 22, Springer-Verlag.

Flaschka, H., and McLaughlin, D.W. [1976] *Cannonically Conjugate Variables for the Korteweg-de Vries Equation and the Toda Lattice with Periodic Boundary Conditions*, Prog. Theor. Phys., **55**, 438–456.

Forest, M.G., and McLaughlin, D.W. [1983] *Modulation of Sinh-Gordon and Sine-Gordon Wavetrains*, Studies in Appl. Math., **68**, 11–59.

Friberg, S.R., Machida, S., and Yamamoto, Y. [1992] *Quantum-Nondemolition Measurement of the Photon Number of an Optical Soliton*, Phys. Rev. Lett., **69**, 3165–3168.

Hasegawa, A., and Kodama, Y. [1992] *Theoretical foundation of optical-soliton concept in fibers*; in Progress in Optics XXX (E. Wolf, ed.), North-Holland, Elsevier Science Publishers B.V., Amsterdam.

Guillemin, V., and Sternberg, S. [1983] *The Gelfand-Cetlin system and the quantization of complex Flag Manifolds*, J. Funct. Anal., **52**, 106–128.

Kaup, D.J., and Newell, A.C. [1978] *Solitons as particles, oscillators, and in slowly changing media: a singular perturbation theory*, Proc. R. Soc. London, Ser. A, **361**, 413–446.

Kovacic, G., and Wiggins, S. [1992] *Orbits homoclinic to resonances, with an application to chaos in model of forced and damped sine-Gordon equation*, Physica D., **57**, 185.

Knörrer, H. [1984] *Singular fibres of the momentum mapping for integrable Hamiltonian systems*, J. Reine u. Ang. Math., **355**, 67–107.

Marsden, J.E., Montgomery, R., and Ratiu, T. [1990] *Cartan-Hannay-Berry phases and symmetry*, Contemporary Mathematics, **97**, 279, (1989); see also Mem. AMS, **436**.

McKean, H.P. [1979] *Integrable Systems and Algebraic Curves*, Lecture Notes in Mathematics, Springer-Verlag, Berlin.

———— [1977] *Theta functions, solitons, and singular curves;* in PDE and Geometry, Proc. of Park City Conference (C.I. Byrnes, ed.).

McLaughlin, D.W., and Overman, E.A. [1993] *Whiskered tori for integrable pde's and chaotic behavior in near integrable pde's*, Surveys in Appl. Math., **1**.

Montgomery, R. [1988] *The connection whose holonomy is the classical adiabatic angles of Hannay and Berry and its generalization to the non-integrable case*, Commun. Math. Phys., **120**, 269–294.

Moser, J. [1980] *Various aspects of integrable Hamiltonian systems*, Progress in Math., **8**, Birkhauser, Boston.

Moser, J., and Veselov, A.P. [1991] *Discrete versions of some classical integrable systems and factorization of matrix polynomials*, Commun. Math. Phys., **139**, 217–243.

Mumford, D. [1983] *Tata Lectures on Theta I and II*, Progress in Math., **28** and **43**, Birkhauser, Boston.

Previato, E. [1985] *Hyperelliptic quasi-periodic and soliton solutions of the nonlinear Schrödinger equation*, Duke Math. Journal, **52**, 329–377.

Ruijsenaars, S.N.M. [1988] *Action-angle maps and scattering theory for some finite-dimensional integrable systems I. The pure soliton case*, Commun. Math. Phys., **115**, 127–165.

Toda, M. [1989] *Theory of Nonlinear Lattices*, Springer-Verlag, New York.

Venakides, S. [1985] *The generation of modulated wavetrains in the solution of the Korteweg-de Vries equation*, Comm. Pure and Appl. Math., **38**, 883–909.

Weinstein, A. [1990] *Connections of Berry and Hannay type for moving Lagrangian submanifolds*, Adv. in Math., **82**, 133–159.

Fields Institute Communications
Volume **3**, 1994

Schur Flows for Orthogonal Hessenberg Matrices

Gregory S. Ammar*
Department of Mathematical Sciences
Northern Illinois University
DeKalb, IL, USA
60115

William B. Gragg[†]
Department of Mathematics
Naval Postgraduate School
Monterey, CA, USA
93943

Abstract. We consider a standard matrix flow on the set of unitary upper Hessenberg matrices with nonnegative subdiagonal elements. The Schur parametrization of this set of matrices leads to ordinary differential equations for the weights and the parameters that are analogous with the Toda flow as identified with a flow on Jacobi matrices. We derive explicit differential equations for the flow on the Schur parameters of orthogonal Hessenberg matrices. We also outline an efficient procedure for computing the solution of Jacobi flows and Schur flows.

1 Introduction

Let $\mathcal{H}_n$ denote the set of unitary upper Hessenberg matrices with nonnegative subdiagonal elements. These matrices bear many similarities with real symmetric tridiagonal matrices, both in terms of their structure, their underlying connections with orthogonal polynomials, and the existence of efficient algorithms for solving

---

1991 *Mathematics Subject Classification*. Primary 15A18; Secondary 34A30.

*The work of the first author was supported in part by the Sonderforchungsbereich 343, Diskrete Strukturen in der Mathematik, Universität Bielefeld. (Internet: ammar@math.niu.edu.)

†The work of the second author was supported in part by a direct grant from the Naval Postgraduate school, by G.S. Magnusson's fund of the Royal Swedish Academy of Sciences, by the Sonderforchungsbereich 343, Diskrete Strukturen in der Mathematik, Universität Bielefeld, and by the Interdisciplinary Project Center for Supercomputing of the ETH, Zürich. William B. Gragg thanks the Department of Mathematics, Naval Postgraduate School, Monterey, CA. Internet: gragg@guinness.math.nps.navy.mil.

Supported in part by the Ministry of Colleges and Universities of Ontario and the Natural Sciences and Engineering Research Council of Canada while visiting The Fields Institute.

eigenproblems for these matrices. The *Schur parametrization* of $\mathcal{H}_n$, which is described below, provides the means for the development of efficient algorithms for this class of matrices.

In this paper we consider a shiftless QR flow on unitary Hessenberg matrices, and derive formulas analogous with those considered in Deift, Nanda and Tomei [1983] for the Toda flow. In particular, we will see that the Schur parameters satisfy a set of ordinary differential equations. We refer to the resulting flow as a *Schur flow*. Explicit differential equations for the Schur parameters are derived for the flow on orthogonal Hessenberg matrices, from which we can conclude that the Schur flow for orthogonal Hessenberg matrices can be regarded as a spatial discretization of the modified Korteweg-de Vries equation. We conclude with some remarks on the computation of the solutions of Toda flows and Schur flows.

2 Unitary Hessenberg matrices

The following proposition gives a parametrization of $\mathcal{H}_n$ that is fundamental in the development of structure-preserving algorithms for eigenproblems for these matrices. For a complex number $|\alpha| \leq 1$, let $G_j(\alpha)$ denote the $n \times n$ unitary transformation in the $(j, j+1)$ coordinate plane given by

$$G_j(\alpha) = \begin{bmatrix} I_{j-1} & & & \\ & -\alpha & \beta & \\ & \beta & \overline{\alpha} & \\ & & & I_{n-j-1} \end{bmatrix},$$

where $\beta := \sqrt{1 - |\alpha|^2} \geq 0$ and I_j denotes the identity matrix of order j. Also let $\tilde{G}_n(\alpha)$ denote the $n \times n$ diagonal matrix

$$\tilde{G}_n(\alpha) = \text{diag}[1, 1, \ldots, 1, -\alpha].$$

Proposition 2.1 *Any* $H \in \mathcal{H}_n$ *can be uniquely expressed as the product*

$$H = G_1(\alpha_1)G_2(\alpha_2) \cdots G_{n-1}(\alpha_{n-1})\tilde{G}_n(\alpha_n),$$

where $|\alpha_j| \leq 1$ *for* $j = 1, \ldots, n-1$ *and* $|\alpha_n| = 1$.

Proof Let $H = [\eta_{jk}]_{j,k=1}^n \in \mathcal{H}_n$. Set $\alpha_1 := -\eta_{11}$ and $\beta_1 := \eta_{21}$. Then by definition, $\beta_1 \geq 0$ and $|\alpha_1|^2 + \beta_1^2 = 1$. Moreover,

$$G_1^H(\alpha_1)H = \begin{bmatrix} 1 & 0 \ 0 \ldots 0 \\ 0 & \\ 0 & H_{n-1} \\ \vdots & \\ 0 & \end{bmatrix},$$

where $H_{n-1} \in \mathcal{H}_{n-1}$ and where G_1^H denotes the complex conjugate transpose of the matrix G_1. Proceeding in this manner, we obtain the unitary matrix $G_{n-1}^H(\alpha_{n-1})$ $\cdots G_2^H(\alpha_2)G_1^H(\alpha_1)H = \text{diag}[1, 1, \ldots, 1, -\alpha_n] = \tilde{G}_n(\alpha_n)$. $\quad \square$

Thus, $H \in \mathcal{H}_n$ is determined by the $2n - 1$ real parameters that compose the *Schur parameters* $\{\alpha_j\}_{j=1}^n$ of H, and we write $H = H(\alpha_1, \ldots, \alpha_n)$. The implicitly determined quantities β_j are called the *complementary Schur parameters* of H.

Although the complementary parameters are mathematically redundant, they are often retained for computational purposes.

The 5×5 unitary Hessenberg matrix $H(\alpha_1, \ldots, \alpha_5)$ is explicitly given below.

$$H(\alpha_1, \ldots, \alpha_5) = \begin{bmatrix} -\alpha_1 & -\beta_1\alpha_2 & -\beta_1\beta_2\alpha_3 & -\beta_1\beta_2\beta_3\alpha_4 & -\beta_1\beta_2\beta_3\beta_4\alpha_5 \\ \beta_1 & -\overline{\alpha}_1\alpha_2 & -\overline{\alpha}_1\beta_2\alpha_3 & -\overline{\alpha}_1\beta_2\beta_3\alpha_4 & -\overline{\alpha}_1\beta_2\beta_3\beta_4\alpha_5 \\ 0 & \beta_2 & -\overline{\alpha}_2\alpha_3 & -\overline{\alpha}_2\beta_3\alpha_4 & -\overline{\alpha}_2\beta_3\beta_4\alpha_5 \\ 0 & 0 & \beta_3 & -\overline{\alpha}_3\alpha_4 & -\overline{\alpha}_3\beta_4\alpha_5 \\ 0 & 0 & 0 & \beta_4 & -\overline{\alpha}_4\alpha_5 \end{bmatrix}.$$

In general, the entries of $H(\alpha_1, \ldots, \alpha_n) = [\eta_{j,k}]_{j,k=1}^n$ are given by

$$\eta_{jk} = \begin{cases} -\overline{\alpha}_{j-1}\beta_j\beta_{j+1}\cdots\beta_{k-1}\alpha_k & \text{for } j < k \\ -\overline{\alpha}_{j-1}\alpha_j & \text{for } j = k \\ \beta_j & \text{for } j = k+1 \\ 0 & \text{for } j > k+1 \end{cases} \tag{2.1}$$

where $\alpha_0 \equiv 1$.

Unitary Hessenberg matrices are fundamentally connected with Szegő polynomials; i.e., with polynomials orthogonal with respect to a measure on the unit circle in the complex plane. In particular, the Schur parameters of a unitary Hessenberg matrix are the recurrence coefficients of the Szegő polynomials determined by a discrete measure on the unit circle.

More specifically, consider the discrete inner product on the unit circle,

$$< f(\lambda), g(\lambda) > := \sum_{j=1}^n \overline{f(\lambda_j)} g(\lambda_j) \omega_j^2, \tag{2.2}$$

where $\{\lambda_j\}_{j=1}^n$ are distinct nodes with $|\lambda_j| = 1$, ω_j^2 is the *Gaussian weight* associated with λ_j, and the bar denotes complex conjugation. The monic polynomials $\chi_j(\lambda)$ that are orthogonal with respect to (2.2) are the monic *Szegő polynomials* associated with the inner product. These polynomials satisfy the *Szegő recurrence relations*

$$\chi_{j+1} = \lambda\chi_j + \alpha_{j+1}\tilde{\chi}_j, \tag{2.3}$$
$$\tilde{\chi}_{j+1} = \overline{\alpha}_{j+1}\lambda\chi_j + \tilde{\chi}_j, \quad j = 0, 1, \ldots, n-1, \tag{2.4}$$

where

$$\chi_0 \equiv \tilde{\chi}_0 \equiv 1,$$
$$\alpha_{j+1} := -<1, \lambda\chi_j>/\delta_j,$$
$$\delta_j := \delta_{j-1}\beta_j^2; \delta_0 = <1, 1>$$
$$\beta_j^2 := 1 - |\alpha_j|^2.$$

It can be shown (Gragg [1982]) that $\chi_j(\lambda) = \det(\lambda I_j - H_j)$, where $H_j = H(\alpha_1, \ldots, \alpha_j)$ is the leading principal submatrix of $H = H(\alpha_1, \ldots, \alpha_n)$ of order j. Moreover, the nodes of the inner product are the eigenvalues of H, while the weight ω_j^2 is the squared modulus of the first component of the eigenvector corresponding to the eigenvalue λ_j. In fact, the nodes and weights uniquely determine $H(\alpha_1, \ldots, \alpha_n)$. This result for *inverse eigenproblem* holds, more generally, for any *normal* Hessenberg matrix with positive subdiagonal elements.

Efficient algorithms have been designed for finding eigenvalues and eigenvectors of unitary Hessenberg matrices using their Schur parametrizations.

These algorithms include the QR algorithm for unitary Hessenberg matrices (Gragg [1986]), an algorithm for solving the orthogonal eigenproblem using two half-size singular value decompositions (Ammar, Gragg and Reichel [1986]), a divide-and-conquer method (Gragg and Reichel [1990]), an approach based on matrix pencils (Bunse-Gerstner and Elsner [1991]), and a unitary analog of the Sturm sequence method (Bunse-Gerstner and He (preprint)). Aspects of inverse eigenproblems for unitary Hessenberg matrices are considered in Ammar, Gragg and Reichel [1991] and efficient algorithms for constructing a unitary Hessenberg matrix from spectral data are presented in Wang [1988], Ammar, Gragg and Reichel [1991], and Reichel, Ammar and Gragg [1991]. Algorithms for eigenproblems and inverse eigenproblems for unitary Hessenberg matrices are useful in several computational problems involving Szegő polynomials, including frequency estimation, least-squares approximation by trigonometric polynomials, procedures for updating and downdating discrete Fourier transforms, the finite trigonometric moment problem and the construction of Gaussian quadrature rules on the unit circle (Gragg [1982], Ammar, Gragg and Reichel [1987], Reichel, Ammar and Gragg [1991], Ammar, Gragg and Reichel [1991]).

3 Schur flows

Let us now consider the well-known flow on $n \times n$ matrices given by the Lax differential equation

$$\dot{H}(t) = \frac{d}{dt}H(t) = [H(t), S(H(t))], \quad H(0) = H_0 \in \mathbb{C}^{n \times n}. \tag{3.1}$$

Here $[H, S] = HS - SH$ is the commutator product of H and S, and $S(H)$ denotes the unique skew-Hermitian matrix such that $H - S(H)$ is upper triangular with real diagonal entries. As is well known (see, e.g., Watkins [1984]), the solution to (3.1) is

$$H(t) = Q^H(t)H_0Q(t), \tag{3.2}$$

where the unitary matrix $Q(t)$ is determined by the *unique* QR factorization $Q(t)R(t) := e^{H_0t}$, where the diagonal elements of the upper triangular matrix $R(t)$ are positive. It follows immediately that the flow is *isospectral*; i.e., the eigenvalues of H(t) are the same as those of H_0 for any t. In fact, the flow preserves many other properties of H_0. For example, if H_0 is real, upper Hessenberg, Hermitian, unitary, or normal, then so is $H(t)$. The flow also preserves the *departure from normality* of H_0.

If H_0 is a normal matrix with eigenvalues $\lambda_1, \dots, \lambda_n$, then we can write

$$H(t) = U(t)\Lambda U^H(t),$$

where $U(t)$ is a unitary matrix and $\Lambda := \mathrm{diag}[\lambda_1, \dots, \lambda_n]$. If, in addition, H_0 is an upper Hessenberg matrix with positive subdiagonal elements, then $H(t)$ is uniquely determined by Λ and $u(t) := U^H(t)e_1$. In other words, any normal Hessenberg matrix with positive subdiagonal elements is uniquely determined by its eigenvalues and the first components of its normalized eigenvectors. (The eigenvectors are scaled so that each component of $u(t)$ is positive.) This fact forms the basis of the inverse eigenvalue problems considered in Gragg and Harrod [1984] and Ammar, Gragg and Reichel [1991]. We refer to the representation of a normal Hessenberg matrix by its eigenvalues and weights as the *internal parametrization* of normal Hessenberg matrices. In contrast, an *external parametrization* would be obtained directly from the entries of the matrix. For real tridiagonal matrices, the external parametrization using the diagonal and subdiagonal entries is obvious. The Schur parametrization provides an external parametrization of unitary Hessenberg matrices. It is the existence of the external parametrization that makes the development of efficient algorithms possible.

If H_0 is an irreducible normal Hessenberg matrix, then the solution $H(t)$ of the flow (3.1) can be represented by the *Gaussian weights* $w(t)$ corresponding to $H(t)$. These weights are given by $w(t) = \overline{u}(t) \circ u(t)$, where $u(t) := U^H(t)e_1$, and the circle denotes the componentwise (Hadamard) product. Direct calculation shows that the vector $w(t)$ satisfies

$$\dot{w} = 2(I - we^T)(\text{Re}\Lambda)w$$

with solution

$$w(t) = \frac{\exp(2\text{Re}\Lambda t)w(0)}{\|\exp(2\text{Re}\Lambda t)w(0)\|_2}. \tag{3.3}$$

This formula, which is valid for any irreducible normal Hessenberg matrix, is utilized in Chu [1984] to analyze the asymptotic behavior of flows on normal Hessenberg matrices.

The following result gives equations for the subdiagonal elements in any Hessenberg flow.

Proposition 3.1 *Let* $H = H(t) = [\eta_{j,k}]_{j,k=1}^n$ *be the solution of the flow (3.1), where* H_0 *is an upper Hessenberg matrix. Then the subdiagonal elements* $\beta_j :=$ $\eta_{j+1,j}$ *of* H *satisfy the differential equations*

$$\dot{\eta}_{j+1,j} = \eta_{j+1,j}(\text{Re}(\eta_{j+1,j+1}) - \text{Re}(\eta_{j,j})).$$

Of course, if $H_0 \in \mathcal{H}_n$, then $H(t) \in \mathcal{H}_n$ for all t; consequently, the flow on $\mathcal{H}_n$ can be regarded as a flow on the Schur parameter pairs $\{(\alpha_j(t), \beta_j(t))\}_{j=1}^n$ of $H(t)$. The following result gives explicit differential equations for the Schur parameters of the flow (3.1) in the case that $H_0 \in \mathcal{H}_n$ is real (i.e., orthogonal). We refer to these equations as the *Schur flow for orthogonal Hessenberg matrices.*

Theorem 3.2 *Let* $H_0 = H(\alpha_1(0), \dots, \alpha_n(0)) \in \mathcal{H}_n$ *with each* $\alpha_j \in \mathbb{R}$. *Then the Schur parameters* $\alpha_k(t)$ *and complementary Schur parameters* $\beta_k(t)$, $k = 1, \dots, n-1$ *of* $H(t)$ *satisfy*

$$\begin{aligned} \dot{\alpha}_k &= \beta_k^2(\alpha_{k+1} - \alpha_{k-1}) \\ \dot{\beta}_k &= \alpha_k\beta_k(\alpha_{k-1} - \alpha_{k+1}), \end{aligned}$$

or equivalently,

$$\dot{\alpha}_k = (1 - \alpha_k^2)(\alpha_{k+1} - \alpha_{k-1}), \tag{3.4}$$

with $\alpha_0 \equiv 1$ *and* $\alpha_n \equiv \alpha_n(0)$.

The formula for $\dot{\beta}_k$ follows from Proposition (3.1) and (2.1). The formula for $\dot{\alpha}_k$ then follows by differentiating the relationship $\alpha_k^2 + \beta_k^2 \equiv 1$. $\qquad\square$

The convergence of the Schur flow depends on the moduli of the eigenvalues of the matrix $\exp(H_0)$. It is shown in Chu [1984] that if H_0 is a real normal Hessenberg matrix with positive subdiagonal elements, and if the eigenvalues of H_0 have distinct real parts, except for complex conjugate pairs, then the matrix flow (3.1) will converge elementwise. In our case, H_0 is an irreducible orthogonal Hessenberg matrix. Its eigenvalues are therefore distinct, of unit modulus, and occur in complex conjugate pairs. It therefore follows that the solution $H(t)$ of (3.1) will converge (elementwise) to a block diagonal matrix, with 2×2 diagonal blocks corresponding to conjugate eigenvalues of H_0, and possibly one or two 1×1 blocks corresponding to real eigenvalues ± 1. Moreover, the real eigenvalues of H_0 can be determined from the sign of α_n and the parity of n (Ammar, Gragg and Reichel [1986]). Thus, the Schur flow on an irreducible orthogonal Hessenberg matrix converges.

4 Computation of Jacobi and Schur flows

Often the use of flows is proposed for finding the eigenvalues of a tridiagonal matrix. However, existing discrete methods are quite fast. For example, the tridiagonal QR method using the Wilkinson shifting strategy is globally quadratically convergent, and almost always cubically convergent. It is shown in Wang [1988] that the unitary Hessenberg QR algorithm is globally cubically convergent with a particular shifting strategy. Moreover, efficient algorithms exist for constructing a Jacobi matrix or a unitary Hessenberg matrix from its eigenvalues and weights (Gragg and Harrod [1984], Ammar, Gragg and Reichel [1991]). The existence of efficient algorithms for solving eigenproblems and inverse eigenproblems for real tridiagonal matrices and unitary Hessenberg matrices leads us to propose the use of discrete methods to compute the solution of the Jacobi flow and Schur flow at any time. In fact, this procedure applies more generally, and less efficiently, to the computation of the solution of the flow (3.1) on the set of normal Hessenberg matrices with positive subdiagonal elements.

Our procedure for computing the solution $H(t_0)$ of the flow (3.1) at any time t_0, where $H(0) = H_0$ is a normal Hessenberg matrix with positive subdiagonal elements, can be divided into two steps.

- **Step 1:** Compute the eigenvalues $\Lambda = \mathrm{diag}[\lambda_1, \ldots, \lambda_n]$ and weights w_0 of H_0.

- **Step 2:** Compute the weights $w(t_0)$ by (3.3), then construct the matrix $H(t_0)$ by solving the inverse eigenproblem.

Observe that if the solution $H(t)$ is desired at several instants of time, Step 1 need only be performed once, and Step 2 can be performed independently and in parallel for each desired value of t.

If H_0 is a Jacobi matrix, Step 1 can be efficiently performed using the adaptation of the QR algorithm described in Golub and Welsch [1969] or with a divide-and-conquer algorithm (Dongarra and Sorensen [1987]), while Step 2 is efficiently performed using the algorithm described in Gragg and Harrod [1984]. If $H_0 \in \mathcal{H}_n$, Step 1 can be performed using an adaptation of the algorithm of Gragg [1986] or with the divide-and-conquer algorithm presented in Gragg and Reichel [1990], while Step 2 can be performed with the inverse unitary Hessenberg QR algorithm presented in Ammar, Gragg and Reichel [1991]. In these two cases, each step of the algorithm requires $O(n^2)$ arithmetic operations. Similar procedures can be used for a general normal Hessenberg matrix H_0, although in this case each step will require

$O(n^3)$ operations since an external parametrization for normal Hessenberg matrices is not known. Nevertheless, this approach is likely to be more efficient than algorithms that solve the differential equations directly.

5 Concluding remarks

We have seen that the standard QR flow on a unitary Hessenberg matrix gives rise to a flow on the corresponding Schur parameters. In the special case of orthogonal Hessenberg matrices, explicit differential equations for the Schur parameters can be obtained. These flows converge. The form of the Schur flow is close to that of the Jacobi flow that arises from the Toda flow, which leads to the question of what other properties of Jacobi flows are shared by Schur flows.

One connection between Jacobi flows and Schur flows is provided by the analogous roles of Jacobi matrices in the study of polynomials orthogonal on an interval of the real axis and unitary Hessenberg matrices in the study of polynomials orthogonal on the unit circle. The study of the Korteweg-de Vries (KdV) partial differential equation provides another connection. While a Jacobi (Toda) flow can be regarded as a spatial discretization of the KdV equation (Ablowitz and Segur [1981]), the Schur flow (3.4) can be shown to be a spatial discretization of a PDE known as the modified KdV (mKdV) equation (see Ablowitz and Segur [1981], p. 121). See Ablowitz and Segur [1981] and Miura [1976] for more on the relationship between the KdV and mKdV equations.

While the Schur parametrization of $\mathcal{H}_n$ provides for the development of efficient algorithms for eigenproblems, the numerical implications of the parametrization, e.g., the conditioning of the maps between $\mathcal{H}_n$ and sets of Schur parameters, are at present not fully understood. Moreover, other external parametrizations of $\mathcal{H}_n$ exist. A further study of the geometry of flows on $\mathcal{H}_n$ may lead to a better understanding of error propagation in algorithms, and may also indicate other parametrizations that may be more natural from a geometric viewpoint. In this regard, ideas presented in Deift, Demmel, Li and Tomei [1991] and Deift, Li and Tomei [1989] may be useful.

References

Ablowitz, M.J., and Segur, H. [1981] *Solitons and the Inverse Scattering Transform,* Soc. Industrial Appl. Math., Philadelphia.

Ammar, G.S., Gragg, W.B., and Reichel, L. [1986] *On the eigenproblem for orthogonal matrices;* in: Proceedings of the 25th Conference on Decision and Control, IEEE, New York, 1063–1066.

———[1987] *Determination of Pisarenko frequency estimates as eigenvalues of an orthogonal matrix;* in: Advanced Algorithms and Architectures for Signal Processing II (F.T. Luk, ed.) Proc. SPIE, **826**, 143–145.

———[1991] *Constructing a unitary Hessenberg matrix from spectral data,* Numerical Linear Algebra; in: Digital Signal Processing and Parallel Algorithms, (G.H. Golub and P. Van Dooren, eds.), Springer, New York, 385–396.

———[1992] *Downdating of Szegő polynomials and data-fitting applications,* Lin. Alg. Appl., **172**, 315–336.

Bunse-Gerstner, A., and Elsner, L. [1991] *Schur parameter pencils for the solution of the unitary eigenproblem,* Lin. Alg. Appl., **154–156,** 741–778.

Bunse-Gerstner, A., and He, C. *A Sturm sequence of polynomials for unitary Hessenberg matrices,* preprint.

Chu, M.T. [1984] *On the global convergence of the Toda lattice for real normal matrices and its applications to the eigenvalue problem,* SIAM J. Math. Anal., **15,** 98–104.

Deift, P., Nanda, T., and Tomei, C. [1983] *Ordinary differential equations and the symmetric eigenvalue problem,* SIAM J. Numer. Anal., **20,** 1–22.

Deift, P., Li, L.-C., and Tomei, C. [1989] *Matrix factorizations and integrable systems,* Comm. Pure Appl. Math., **42,** 443–521.

Deift, P., Demmel, J., Li, L.-C., and Tomei, C. [1991] *The bidiagonal singular value problem and Hamiltonian mechanics,* SIAM J. Numer. Anal., **28,** 1463–1516.

Dongarra, J.J., and Sorensen, D.C. [1987] *A fully parallel algorithm for the symmetric eigenvalue problem,* SIAM J. Sci. Stat. Comput. **8,** s139–s154.

Golub, G.H., and Welsch, J.H. [1969] *Calculation of Gauss quadrature rules,* Math. Comp., **23,** 221–230.

Gragg, W.B. [1982] *Positive definite Toeplitz matrices, the Arnoldi process for isometric operators, and Gaussian quadrature on the unit circle (in Russian);* in: Numerical Methods in Linear Algebra (E.S. Nikolaev, ed.), Moscow University Press, Moscow, 16-32.

———[1986] *The QR algorithm for unitary Hessenberg matrices.* J. Comput. Appl. Math., **16,** 1–8.

Gragg, W.B., and Harrod, W.J. [1984] *The numerically stable reconstruction of Jacobi matrices from spectral data,* Numer. Math., **44,** 317–335.

Gragg, W.B., and Reichel, L. [1990] *A divide and conquer method for unitary and orthogonal eigenproblems,* Numer. Math., **57,** 695–718.

Miura, R.M. [1976] *The Korteweg-de Vries equation: A survey of results,* SIAM Rev., **18,** 412–459.

Reichel, L., Ammar, G.S., and Gragg, W.B. [1991] *Discrete least squares approximation by trigonometric polynomials,* Math. Comp., **57,** 273–289.

Wang, T.-L. [1988] *Convergence of the QR Algorithm with Origin Shifts for Real Symmetric Tridiagonal and Unitary Hessenberg Matrices,* Ph.D. Dissertation, Dept. of Mathematics, University of Kentucky, Lexington, KY.

Watkins, D.S. [1984] *Isospectral Flows,* SIAM Rev. **26,** 379–392.

Fields Institute Communications
Volume **3**, 1994

Sub-Riemannian Optimal Control Problems

Anthony M. Bloch[*]
Department of Mathematics
The Ohio State University
Columbus, OH, USA
43210

Peter E. Crouch[†]
Center for Systems Science and Engineering
Arizona State University
Tempe, AZ, USA
85287

Tudor S. Ratiu[‡]
Department of Mathematics
University of California
Santa Cruz, CA, USA
95064

Abstract. In this paper we analyze the Hamiltonian structure of certain kinematic optimal control problems. In particular we consider the so-called sub-Riemannian optimal control problems (see Brockett [1973, 1982], Baillieul [1975, 1978]) which have a smaller number of control vector fields than the dimension of the state space. We analyze a class of systems defined on compact semisimple Lie groups, in which we use the structure of a Riemannian symmetric space. The extremals are shown to be Lie-Poisson equations for the singular generalized rigid body. We consider the singular equations as the limit of the full generalized rigid body equations and compare the sub-Riemannian problem to the singular systems of Fomenko [1982]. This paper consists partly of a survey and partly of some new approaches to these problems.

1 Introduction

In this paper we consider a number of aspects of the optimal control of singular kinematic control systems on Riemannian manifolds. This problem has an interest-

1991 *Mathematics Subject Classification.* Primary 54C40, 54E20; Secondary 46E25, 20C20.

[*]Partially supported by NSF Grants DMS-9002136 and PYI DMS-9157566, AFOSR Grant F49620-93-1-0037

[†]Partially supported by NSF Grants DMS-91011906 and INT-88-14643

[‡] Partially supported by NSF Grant DMS-9122708

Supported in part by the Ministry of Colleges and Universities of Ontario and the Natural Sciences and Engineering Research Council of Canada while visiting The Fields Institute.

ing history (see Brockett [1973, 1982], Baillieul [1975, 1978], Günther [1982]) and has also received much interest of late because of its importance in the kinematic (i.e. velocity) control of systems with nonholonomic constraints (see Bloch and Crouch [1992], Brockett and Dai [1992], Krishnaprasad and Yang [1991], Lafferiere and Sussmann [1991], Montgomery [1990], Murray and Sastry [1990] and others). (See also the work of Strichartz [1986, 1987].) Our goal here is to recall some of the results in this area, to give a variational derivation of the geodesics, and to discuss their Hamiltonian structure in some special cases. In particular, we consider the singular optimal control problem connected to Riemannian symmetric spaces, in which case we show that the optimal trajectories satisfy the Lie-Poisson equations for a singular generalized rigid body. These are generalized rigid body equations whose kinetic energy is singular. We remark that these equations are quite different from those obtained by letting a component of the inertia tensor tend to zero in the equations. Fomenko [1982] also examined singular rigid body equations on symmetric spaces in a rather different sense. However, we do show that by using a general formalism of Mishchenko and Fomenko [1982], the singular generalized rigid body equations may be obtained from the full generalized rigid body equations by a suitable limiting procedure. Baillieul [1975] analyzed the singular optimal control problem on $SO(3)$ when the kinetic energy metric equals the ambient metric (i.e. the "inertia tensor" is the identity). We of course recover his equations in this case.

In the case of $SO(3)$ our equations are still integrable, but different from the integrable cases of Faybusovich [1988], who examined integrability of optimal trajectories for certain nonsingular control problems.

We remark also that a procedure for determining the optimal trajectories via the maximum principle for control systems on Lie groups is discussed in Brockett [1973] and Jurdjevic [1991a, c]. Our analysis represents an alternative to this approach.

The contents of the paper are as follows: in section 2 we discuss the general optimal control problem on a Riemannian manifold; in section 3 we consider the case of a system on a compact semisimple Lie group; in section 4 we specialize to the case where we have the structure of a symmetric space, and discuss the Hamiltonian nature of the necessary conditions; in section 5 we show how to obtain the singular generalized rigid body equations as a limit of the Mishchenko-Fomenko formalism and we discuss briefly Fomenko's (non sub-Riemannian) singular rigid bodies.

2 The kinematic sub-Riemannian control problem

Let M^n be a Riemannian manifold of dimension n with metric denoted by $\langle \cdot, \cdot \rangle$. The corresponding Riemannian connection and covariant derivative will be denoted by ∇ and $D/\partial t$ respectively. Now assume that M is such that there exist smooth vector fields $X^1(q), \cdots, X^n(q)$ satisfying $\langle X^i(q), X^j(q) \rangle = \delta_{ij}$, an orthonormal frame for $T_q M$ for all $q \in M$. This of course limits the class of manifolds we consider, but is satisfied for the main case of interest to us, M a Lie group G.

We now define the kinematic control system on M

$$\frac{dq}{dt} = \sum_{i=1}^{m} u_i X^i(q), \quad m < n. \tag{2.1}$$

The singular optimal control problem for (2.1) is defined by

$$\min_u \int_0^T \frac{1}{2} \sum_{i=1}^m u_i^2(t)\,dt; \quad q(0) = q_0 \quad q(T) = q_T \tag{2.2}$$

subject to (2.1).

This may be posed as a variational problem on M as follows:

Define the constraints

$$\omega_k\left(\frac{dq}{dt}\right) = \langle X^k, \frac{dq}{dt}\rangle = 0, \quad m < k \leq n \tag{2.3}$$

and let

$$Z_t = \sum_{k=m+1}^n \lambda_k(t) X^k \tag{2.4}$$

where the λ_k are Lagrange multipliers. By virtue of the orthonormality of the X^i the optimal control problem then becomes

$$\min_q J(q) = \min_q \int_0^T \left(\frac{1}{2}\langle \frac{dq}{dt}, \frac{dq}{dt}\rangle + \langle Z_t, \frac{dq}{dt}\rangle\right) dt$$

$$\langle Z_t, \frac{dq}{dt}\rangle = 0. \tag{2.5}$$

We now briefly derive the necessary conditions for the regular extremals of this variational problem following Milnor [1963] and Crouch and Silva-Leite [1991]. (For interesting recent work on abnormal extremals see for example Bryant and Hu [1993], Montgomery [1992], and Sussmann [1992].)

The tangent space to the space Ω of C^2 curves satisfying the boundary conditions of (2.2) is denoted by $T_q\Omega$. It is the space of C^1 vector fields $t \to W_t$ along $q(t)$ satisfying $W_0 = 0 = W_T$. The curve $t \to \frac{DW_t}{\partial t}$ in TM is continuous. Exponentiating a vector field in $T_q\Omega$ we obtain a one-parameter variation of q:

$$\alpha : [0, T] \times (-\epsilon, \epsilon) \to M,$$

$$\alpha_u(t) = \alpha(t, u) = \exp_{q(t)}(uW_t),$$

where exp is the exponential mapping on M. Note that $\alpha_u(0) = q(0) = q_0$, $\alpha_u(T) = q(T) = q_T$, $\alpha_0(t) = q(t)$, $\frac{\partial \alpha_0(t)}{\partial u} = W_t$, $0 \leq t \leq T$.

The necessary conditions for regular extremals are obtained from

$$\frac{d}{du} J(\alpha_u)\,|_{u=0} = 0, \tag{2.6}$$

where

$$J(\alpha_u) = \int_0^T \left(\frac{1}{2}\langle \frac{\partial \alpha_u}{\partial t}, \frac{\partial \alpha_u}{\partial t}\rangle + \langle Z_t(\alpha_u), \frac{\partial \alpha_u}{\partial t}\rangle\right) dt.$$

Now

$$\left.\frac{DJ(\alpha_u)}{du}\right|_{u=0} = \int_0^T \left(\langle\frac{dq}{dt}, \frac{DW_t}{\partial t}\rangle + \langle\nabla_{W_t}Z_t, \frac{dq}{dt}\rangle\right.$$
$$\left. + \langle Z_t, \frac{D}{\partial t}W_t\rangle\right) dt$$

$$= \int_0^T \left(-\langle\frac{D}{dt}V_t, W_t\rangle - \langle\frac{D}{\partial t}Z_t, W_t\rangle\right.$$
$$\left. - \langle\nabla_{Z_t}V_t, W_t\rangle + \langle[W_t, Z_t], V_t\rangle\right] dt \tag{2.7}$$

where $V_t = \frac{dq}{dt} = \sum_{i=1}^m v_i(t)X^i(q)$. Here we use $\nabla_W Z = \nabla_Z W + [W, Z]$ and $Z[\langle V, W\rangle] = \langle\nabla_Z V, W\rangle + \langle V, \nabla_Z W\rangle$.

3 The necessary conditions on a compact semisimple lie group

Now let $M = G$, G a compact semisimple Lie group, with Lie algebra $\mathfrak{g}$, and let $\langle\langle\cdot, \cdot\rangle\rangle = -\frac{1}{2}\kappa(\cdot, \cdot)$ where κ is the Killing form on $\mathfrak{g}$.

Let J be a positive definite linear mapping $J : \mathfrak{g} \to \mathfrak{g}$ satisfying

$$\langle\langle JX, Y\rangle\rangle = \langle\langle X, JY\rangle\rangle$$

$$\langle\langle JX, X\rangle\rangle \geq 0 \quad (= 0 \text{ if and only if } X = 0).$$

Then if $X, Y \in \mathfrak{g}$, R_g is right translation on G by $g \in G$, and $X_g^r = X^r(g) = R_{g*}X$ and $Y_g^r = Y^r(g) = R_{g*}Y$ are corresponding right invariant vector fields,

$$\langle X^r(g), Y^r(g)\rangle = \langle\langle X, JY\rangle\rangle \tag{3.1}$$

defines a right-invariant metric on G. Corresponding to the right invariant metric $\langle\ ,\ \rangle$ there is a unique Riemannian connection ∇ (see e.g. Nomizu [1956]). ∇ defines a bilinear form on $\mathfrak{g}$:

$$(X, Y) \to \nabla_X Y = \frac{1}{2}\{[X, Y] + J^{-1}[X, JY] + J^{-1}[Y, JX]\}, \quad X, Y \in \mathfrak{g}, \tag{3.2}$$

and the expression for ∇ on right-invariant vector fields on G is

$$(\nabla_{X^r}Y^r)(g) = (\nabla_X Y)_g^r. \tag{3.3}$$

Now choose an orthonormal basis e_i on $\mathfrak{g}$, $\langle\langle e_i, Je_j\rangle\rangle = \delta_{ij}$, and extend it to a right invariant orthonormal frame on T_gG, $X^i(g) = R_{g*}e_i \equiv X^{ir}(g)$.

In this setting the computation of $\frac{d}{du}J(\alpha_u)|_{u=0}$ is reduced to a computation in the Lie algebra. Now if in $\mathfrak{g}$

$$V_t = \sum_{i=1}^m v_i(t)e_i, \quad \dot{V}_t = \sum_{i=1}^m \dot{v}_i(t)e_i$$

and similarly for $W_t = \sum_{i=1}^n w_i(t)e_i$ and Z_t, at the group level we have for $\frac{dq}{dt} = V_t^r$,

$$\frac{DV_t^r}{dt} = \frac{D}{dt}\left(\sum_{i=1}^m v_i(t)X^i\right) = (\dot{V}_t + \nabla_{V_t}V_t)_g^r$$

$$= (\dot{V}_t + J^{-1}[V_t, JV_t])_g^r \tag{3.4}$$

by (3.2) and (3.3), and

$$\frac{DZ_t^r}{dt} = (\dot{Z}_t + \nabla_{V_t} Z_t)_g^r. \tag{3.5}$$

Now, by (3.1),

$$\frac{d}{du} J(\alpha_u) \mid_{u=0}$$

$$= \int_0^T (-\langle\langle \dot{V}_t + J^{-1}[V_t, JV_t] + \dot{Z}_t + \nabla_{Vt} Z_t + \nabla_{Z_t} V_t, \; JW_t \rangle\rangle$$

$$+ \langle\langle [W_t, Z_t], JV_t \rangle\rangle) dt. \tag{3.6}$$

Using,

$$\langle\langle [X, Y], Z \rangle\rangle + \langle\langle Y, [X, Z] \rangle\rangle = 0$$

the necessary conditions (2.6) imply that

$$\dot{V}_t + J^{-1}[V_t, JV_t] + \dot{Z}_t + \nabla_{V_t} Z_t + \nabla_{Z_t} V_t + J^{-1}[JV_t, Z_t] = 0. \tag{3.7}$$

By (3.2)

$$\nabla_{V_t} Z_t + \nabla_{Z_t} V_t = J^{-1}[V_t, JZ_t] + J^{-1}[Z_t, JV_t].$$

Hence the necessary conditions on $\mathfrak{g}$ are

$$\dot{V}_t + J^{-1}[V_t, JZ_t] + \dot{Z}_t + J^{-1}[V_t, JV_t] = 0 \tag{3.8}$$

with the constraint

$$\langle \frac{dg}{dt}, Z_t \rangle = \langle\langle V_t, JZ_t \rangle\rangle = 0. \tag{3.9}$$

Equations (3.8) are identical to the equations (3.7) of Brockett [1973] in the case $J = I$. We can see this as follows:

Firstly write the system (2.1) as

$$\frac{dg}{dt} = \sum_{i=1}^m u_i B^i g, \quad m < n, \tag{3.10}$$

where $B_i \in \mathfrak{g}$ and the $X^i(g) = B^i g$ are thus right invariant vector fields on G. Then $V_t = \sum_{i=1}^m u_i(t) B^i$. Now set $L_t = V_t + Z_t$. Then equation (3.8) becomes

$$\dot{L}_t = J^{-1}[JL_t, V_t] \tag{3.11}$$

or

$$\dot{L}_t = \sum_{i=1}^m u_i J^{-1}[JL_t, B^i]. \tag{3.12}$$

Setting $J = I$ we recover precisely Brockett's equations (in the case of zero drift). Note also that in the case $J = I$ our equations (3.8) assume the symmetric form

$$\dot{V}_t + [V_t, Z_t] + \dot{Z}_t = 0 \, . \tag{3.13}$$

Brockett obtained his equations by applying the maximum principle to Lie groups, while we have taken a direct variational approach. The approaches are of course equivalent (see also the next section).

4　A special case

Suppose now that G/K is a Riemannian symmetric space, G as in Section 3, K a closed subgroup of G with Lie algebra $\mathfrak{k}$. Then $\mathfrak{g} = \mathfrak{p} \oplus \mathfrak{k}$ with

$$[\mathfrak{p}, \mathfrak{p}] \subset \mathfrak{k}$$
$$[\mathfrak{p}, \mathfrak{k}] \subset \mathfrak{p} \tag{4.1}$$
$$[\mathfrak{k}, \mathfrak{k}] \subset \mathfrak{k}$$

and $\langle\langle \mathfrak{k}, \mathfrak{p} \rangle\rangle = 0$. We now want to consider the necessary conditions (3.8) in this case.

Assume that $e_1, \ldots, e_m$ is a basis for $\mathfrak{p}$ and $e_{m+1}, \ldots e_n$ is a basis for $\mathfrak{k}$. Suppose that $J : \mathfrak{p} \to \mathfrak{p}$ and $J : \mathfrak{k} \to \mathfrak{k}$. Then $\langle\langle V_t, JZ_t \rangle\rangle = 0$ for $Z_t = \sum_{i=m+1}^{n} \lambda_i(t) e_i \in \mathfrak{k}$ and $V_t = \sum_{k=1}^{m} v_i(t) e_i \in \mathfrak{p}$.

Since $\dot{V}_t + J^{-1}[V_t, JZ_t] \in \mathfrak{p}$ and $\dot{Z}_t + J^{-1}[V_t, JV_t] \in \mathfrak{k}$, the necessary conditions (3.8) become

$$\dot{V}_t = J^{-1}[JZ_t, V_t]$$
$$\dot{Z}_t = J^{-1}[JV_t, V_t] \tag{4.2}$$

or, if $P_t = JV_t$ and $Q_t = JZ_t$

$$\dot{P}_t = [Q_t, J^{-1}P_t]$$
$$\dot{Q}_t = [P_t, J^{-1}P_t]. \tag{4.3}$$

Now we can easily show that equations (4.3) are Hamiltonian with respect to the Lie-Poisson structure on $\mathfrak{g}$.

Recall that for F, H functions on $\mathfrak{g}$, their $(-)$ Lie-Poisson bracket is given by

$$\{F, H\}(X) = -\langle\langle X, [\nabla F(X), \nabla H(X)] \rangle\rangle, \quad X \in \mathfrak{g}, \tag{4.4}$$

where $dF(X) \cdot Y = \langle\langle \nabla F(X), Y \rangle\rangle$.

For $H(X)$ a given Hamiltonian, we thus have the Lie-Poisson equations $\dot{F}(X) = \{F, H\}(X)$. Letting $F(X) = \langle\langle A, X \rangle\rangle$, $A \in \mathfrak{g}$, we obtain

$$\langle\langle A, \dot{X} \rangle\rangle = -\langle\langle X, [A, \nabla H(X)] \rangle\rangle$$
$$= \langle\langle A, [X, \nabla H(X)] \rangle\rangle$$

and hence

$$\dot{X} = [X, \nabla H(X)]. \tag{4.5}$$

For $H(M) = \frac{1}{2}\langle\langle M, J^{-1}M \rangle\rangle$, $M \in \mathfrak{g}$ and J as in Section 3, we obtain the generalized rigid body equations

$$\dot{M} = [M, J^{-1}M]. \tag{4.6}$$

Now for $X = P + Q \in \mathfrak{p} \oplus \mathfrak{k}$, let $H(X) = H(P) = \frac{1}{2}\langle\langle P, J^{-1}P \rangle\rangle$, $P \in \mathfrak{p}$. Then $\nabla H(X) = J^{-1}P \in \mathfrak{p}$ and equations (4.5) become

$$(Q + P)\dot{} = [Q + P, J^{-1}P],$$

or

$$\dot{P} = [Q, J^{-1}P]$$
$$\dot{Q} = [P, J^{-1}P],$$

precisely equations (4.3).

Thus equations (4.3) are Lie-Poisson with respect to the "singular" Hamiltonian $H(P)$. Summarizing then, we have

Theorem 4.1 *The optimal trajectories for the singular optimal control problem (2.1)–(2.2) on a Riemannian symmetric space are given by equations (4.3). These equations are Lie Poisson with respect to a singular rigid body Hamiltonian on $\mathfrak{g}$.*

We see therefore that we can obtain the singular optimal trajectories by letting $J|_{\mathfrak{k}} \to \infty$ in the full rigid body Hamiltonian $H(X) = \frac{1}{2}\langle\langle X, J^{-1}X\rangle\rangle$, thus obtaining the singular Hamiltonian $H(P) = \frac{1}{2}\langle\langle P, J^{-1}P\rangle\rangle$.

This observation also enables us to obtain the singular rigid body equations directly by a limiting process from the full rigid body equations. The key is the correct choice of angular velocity and momentum variables corresponding to the Lie algebra decomposition $\mathfrak{g} = \mathfrak{p} \oplus \mathfrak{k}$.

In the notation of equation (4.3) we write an arbitrary element of $\mathfrak{g}$ as $M = JV + Q$, $JV \in \mathfrak{p}$, $Q \in \mathfrak{k}$. Then the generalized rigid body equations (4.6) become

$$J\dot{V}_t = [Q_t, V_t] + [JV_t, J^{-1}Q_t]$$
$$\dot{Q}_t = [Q_t, J^{-1}Q_t] + [JV_t, V_t]. \tag{4.7}$$

Letting $J|_{\mathfrak{k}} \to \infty$ we obtain

$$J\dot{V}_t = [Q_t, V_t]$$
$$\dot{Q}_t = [JV_t, V_t]. \tag{4.8}$$

We note that this is a mixture between the Lagrangian and Hamiltonian pictures. While the variables in $\mathfrak{k}$ are momenta (and should really be viewed as lying in $\mathfrak{k}^*$) the variables in $\mathfrak{p}$ are velocities. These variables are not only the natural ones in which to take the limit in the full rigid body equations, but are natural from the point of view of the maximum principle, for the variables Q correspond to the constraints and therefore are naturally viewed as costates.

Note also that this reduction may also be viewed as a generalized Routhian reduction as in Marsden and Scheurle [1992]. The group variables are the only ones which are Legendre transformed. This will be discussed further in a future paper.

As mentioned in the previous section, the necessary conditions above may also be derived directly from the maximum principle developed for Lie groups (see e.g. Brockett [1973], Jurdjevic [1991]). The Hamiltonian in the maximum principle of the system on a Lie group as discussed above is precisely $\frac{1}{2}\langle P_t, J^{-1}P_t\rangle$. This is just the sum of the Hamiltonians corresponding to each of the vector fields X_i.

We note also the following interpretation of the fact that $J|_{\mathfrak{k}} \to \infty$ (due to R. Brockett). Write $M \in \mathfrak{g}$ now as $M = JZ + P$, $Z \in \mathfrak{k}$, $P \in \mathfrak{p}$. Then the Hamiltonian (in the maximum principle) becomes

$$H(M) = \langle M, J^{-1}M\rangle = \langle JZ + P, J^{-1}(JZ + P)\rangle$$
$$= \langle JZ, Z\rangle + \langle P, J^{-1}P\rangle. \tag{4.9}$$

Letting $J\mid_{\mathfrak{k}} \to \infty$ we see that the cost becomes infinite unless $Z = 0$, i.e. the constraints are satisfied.

Example 4.2 The simplest non-trivial example one can consider is the symmetric space $SO(3)/SO(2)$.

In this case $\mathfrak{g} = \mathfrak{k} \oplus \mathfrak{p}$ becomes $so(3) = so(2) \oplus \mathbb{R}^2$. Reflecting this decomposition we may represent matrices in $so(3)$ as

$$\left[\begin{array}{c|cc} 0 & -\omega_3 & \omega_2 \\ \hline \omega_3 & 0 & -\omega_1 \\ -\omega_2 & \omega_1 & 0 \end{array}\right] \tag{4.10}$$

with the lower 2×2 block in $so(2)$.

This example illustrates the importance of writing the optimal equations in the natural variables $M = JV + Q$ in order to understand the limiting process in equations (4.7) and (4.8).

We write

$$JV_t + Q_t = \begin{bmatrix} 0 & -J_3\omega_3 & J_2\omega_2 \\ J_3\omega_3 & 0 & -m_1 \\ -J_2\omega_2 & m_1 & 0 \end{bmatrix}. \tag{4.11}$$

Here $Q_t \in so(2)$ has "momentum" variable m_1.

Then equations (4.8) or

$$(JV_t + Q_t)^{\cdot} = [JV_t + Q_t, V_t] \tag{4.12}$$

become for $\mathfrak{g} = so(3)$

$$\begin{aligned} \dot{m}_1 &= (J_2 - J_3)\omega_2\omega_3 \\ J_2\dot{\omega}_2 &= -m_1\omega_3 \\ J_3\dot{\omega}_3 &= m_1\omega_2. \end{aligned} \tag{4.13}$$

The full rigid body equation in these variables are

$$(JV_t + Q_t)^{\cdot} = [JV_t + Q_t, V_t + J^{-1}Q_t] \tag{4.14}$$

which are for $\mathfrak{g} = so(3)$

$$\begin{aligned} \dot{m}_1 &= (J_2 - J_3)\omega_2\omega_3 \\ J_2\dot{\omega}_2 &= \left(\frac{J_3}{J_1} - 1\right)\omega_3 m_1 \\ J_3\dot{\omega}_3 &= \left(1 - \frac{J_2}{J_1}\right)m_1\omega_2 \end{aligned} \tag{4.15}$$

which clearly limits to (4.13) as $J_1 \to \infty$.

Note that if we write the rigid body equations in the usual form

$$J(V_t + Z_t)^{\cdot} = [J(V_t + Z_t), V_t + Z_t]$$

yielding for $\mathfrak{g} = so(3)$

$$\begin{aligned} J_1\dot{\omega}_1 &= (J_2 - J_3)\omega_2\omega_3 \\ J_2\dot{\omega}_2 &= (J_3 - J_1)\omega_1\omega_3 \\ J_3\dot{\omega}_3 &= (J_1 - J_2)\omega_2\omega_1, \end{aligned} \tag{4.16}$$

we see that the limiting process $J_1 \to \infty$ makes no sense. The same is true for the rigid body in the momentum representation.

We remark that (4.10), despite its singular nature is still integrable, for we still have two conserved quantities, the Hamiltonian $H(\omega) = J_2\omega_2^2 + J_3\omega_3^2 (= \frac{1}{2}\langle P, J^{-1}P \rangle)$ and the Casimir $C(\omega) = m_1^2 + J_2^2\omega_2^2 + J_3^2\omega_2^2$. (Recall that a Casimir function for a Poisson structure is a function that commutes with every other function under the Poisson bracket.) $\square$

It is also interesting to consider the case $J = I$. Equations (4.3) then become

$$\dot{P}_t = [Q_t, P_t] \qquad (4.17)$$
$$\dot{Q}_t = 0.$$

Hence $Q_t = Q$ is constant.

Similarly, considering (4.2), we obtain

$$\dot{V}_t = [Z_t, V_t], \quad Z_t = Z \text{ constant.} \qquad (4.18)$$

This is of course solvable: $V_t = Ad_{e^{Zt}}V_0$ and $u_i(t) = << e_i, Ad_{e^{Zt}}V_0 >>$ (see the equivalent expression (3.5) of Brockett [1973]). Consider again the case $SO(3)/SO(2)$. Since $V_t \in \mathbb{R}^2$ and $Z_t \in so(2)$ we may set

$$Z = \begin{bmatrix} 0 & 0 & 0 \\ 0 & 0 & -\phi \\ 0 & \phi & 0 \end{bmatrix}, \quad \phi \text{ fixed.}$$

Then

$$e^{Zt} = \begin{bmatrix} 1 & 0 & 0 \\ 0 & \cos\phi t & -\sin\phi t \\ 0 & \sin\phi t & \cos\phi t \end{bmatrix}.$$

Hence the optimal evolution of V_t (or equivalently the optimal controls – see (2.1)) is given by rotation. This recovers precisely the result of Baillieul [1975] who indeed analyzed the case $J = I$ in dimension 3. (See Baillieul [1975], page III-5.)

5 Generalized singular rigid bodies

In this section we show how our singular (sub-Riemannian) rigid body equations may be obtained as a limit of Fomenko's generalized rigid body equations. This formulation is useful for the analysis of integrability. Recall (see e.g. Fomenko and Trofimov [1988]) that if $\mathfrak{g}$ is a complex semisimple, split real semisimple, or compact real semisimple Lie algebra, the generalized free rigid body equations are given by $\dot{X} = [X, \varphi_{a,b}X]$, where $\varphi_{a,b} = ad_a^{-1}ad_b$ on the sum of the root spaces for all non-zero roots, a and b are elements of the Cartan algebra, a is in general position, i.e., the value of all non-zero roots on a is never zero, and $\varphi_{a,b}$ restricted to the Cartan algebra is an arbitrary linear map symmetric relative to the Killing form. These equations are Hamiltonian relative to the Lie-Poisson structure on $\mathfrak{g}$ (thought of as $\mathfrak{g}^*$, the identification being given by the Killing form) and Hamiltonian $X \mapsto \frac{1}{2}\langle\langle X, \varphi_{a,b}(X) \rangle\rangle$. They form a completely integrable system on the generic adjoint orbits of $\mathfrak{g}$, the independent integrals in involution being chosen from the Poisson commuting family $X \mapsto f(X + \lambda a)$, for λ an arbitrary number and f an arbitrary invariant function on $\mathfrak{g}$. If $\mathfrak{g}$ is the intersection of the compact and normal real forms of a complex semisimple Lie algebra, the same construction holds with the sole change that a and b are chosen from the Cartan subalgebra of the compact real form or from one of its central extensions. For example, if $\mathfrak{g} = so(n)$, then $b := \text{diag}(i\lambda_1, \ldots, i\lambda_n)$, $\lambda_j \in \mathbb{R}$, $\lambda_i \neq \lambda_j$, $a := -ib^2$, gives the operator $\psi(X) := \varphi_{-ib^2,b}(X)$ whose inverse is $\psi^{-1}(Y) = \Lambda Y + Y\Lambda$, for $\Lambda = \text{diag}(\lambda_1, \ldots, \lambda_n)$. The equations of motion are $\dot{X} = [X, \psi(X)]$ which are the generalization of the free rigid body equations if $\lambda_i + \lambda_j > 0$ for all $i \neq j$. X represents the angular momentum, $\psi(X)$ the angular velocity and Λ the moment of inertia, all in the body representation; the Hamiltonian is the kinetic energy $\frac{1}{2}\langle\langle X, \psi(X)\rangle\rangle$.

We consider here a slight variation of the Fomenko approach for $\mathfrak{g} = so(n)$. We write the generalized rigid body equation in momentum space as $\dot{M} = [M, \varphi_{a,b}M]$ where we take $a = \text{diag}(i\lambda_1, \ldots, i\lambda_n)$, $b = \text{diag}(i\lambda_1^2, \ldots, i\lambda_n^2)$ $(a, b \in u(n))$. Then $\varphi_{a,b}(X) = (x_{ij}(\lambda_i + \lambda_j))$. For the nonsingular rigid body it is assumed that $\lambda_i + \lambda_j \neq 0$ for all i, j. Let E_{ij} be the matrix in $so(n)$ with 1 in the ijth position and zero elsewhere. We have

Proposition 5.1 *The sub-Riemannian singular rigid body equations for* $\mathfrak{g} = so(n)$ *are given by* $\dot{M} = [M, \widehat{\varphi}_{a,b}M]$ *where* $a = diag(i\lambda_1, \ldots i\lambda_n)$, $b = diag(i\lambda_1^2, \ldots i\lambda_n^2)$ *and*

$$\widehat{\varphi}_{a,b}(X) = \lim_{\lambda_i \to -\lambda_j} \varphi_{a,b}(X) \text{ for all } i, j \text{ such that } E_{ij} \in \mathfrak{k}.$$

Note that the practical effect of the limiting process in Proposition 5.1 is to set $\lambda_i + \lambda_j = 0$ for the relevant i, j in the terms $x_{ij}(\lambda_i + \lambda_j)$. Let us illustrate the singular case for $SO(3)$. Let $\lambda_2 \to -\lambda_3$. Then

$$\widehat{\varphi}_{a,b}M = \lim_{\lambda_2 \to -\lambda_3} \varphi_{a,b} \begin{bmatrix} 0 & -m_3 & m_2 \\ m_3 & 0 & -m_1 \\ -m_2 & m_1 & 0 \end{bmatrix}$$

$$= \begin{bmatrix} 0 & -m_3(\lambda_1 + \lambda_2) & m_2(\lambda_1 - \lambda_2) \\ m_3(\lambda_1 + \lambda_2) & 0 & 0 \\ -m_2(\lambda_1 - \lambda_2) & 0 & 0 \end{bmatrix}.$$

Now take the bracket with M:

$$[M, \widehat{\varphi}_{a,b}M] = \begin{bmatrix} 0 & -m_1 m_2(\lambda_1 - \lambda_2) & -m_1 m_3(\lambda_1 + \lambda_2) \\ -m_1 m_2(\lambda_1 - \lambda_2) & 0 & m_2 m_3(-2\lambda_2) \\ m_1 m_3(\lambda_1 + \lambda_2) & m_3 m_2(2\lambda_2) & 0 \end{bmatrix} \tag{5.1}$$

Thus the equations of motion are

$$\begin{aligned} \dot{m}_1 &= 2m_2 m_3 \lambda_2, \\ \dot{m}_2 &= -m_1 m_3(\lambda_1 + \lambda_2), \\ \dot{m}_3 &= m_1 m_2(\lambda_1 - \lambda_2). \end{aligned} \tag{5.2}$$

Now set

$$\lambda_1 + \lambda_2 = \frac{1}{J_3},$$

$$\lambda_1 - \lambda_2 = \frac{1}{J_2}.$$

Hence

$$2\lambda_2 = \frac{J_2 - J_3}{J_2 J_3}.$$

Thus (5.2) becomes

$$\dot{m}_1 = \frac{(J_2 - J_3)}{J_2 J_3} m_2 m_3$$

$$\dot{m}_2 = -\frac{1}{J_3} m_1 m_3 \qquad\qquad (5.3)$$

$$\dot{m}_3 = \frac{1}{J_2} m_1 m_2,$$

precisely equations (4.13) in momentum space.

In the nonsingular case the generalized rigid body equations are known to be integrable (see Manakov [1976], Mishchenko and Fomenko [1976, 1978], Adler and van Moerbeke [1980], Ratiu [1980]). The Manakov approach described above involves, in the case of $\mathfrak{g} = so(n)$, writing the equation as a Lax pair with parameter of the form

$$(M + \mu\Lambda^2)^\cdot = [M + \mu\Lambda^2, \Omega + \mu\Lambda], \qquad\qquad (5.4)$$

$\Lambda = \mathrm{diag}\,(\lambda_1, \ldots, \lambda_n)$, $M = \Lambda\Omega + \Omega\Lambda$. Here Ω and M are the angular velocity and momentum respectively. There are then observed to be sufficient independent integrals in involution among the coefficients of μ in $Tr(M + \mu\Lambda^2)^k$, $k = 2, \ldots n$.

In our variant of the generalized rigid body we can still get a system with parameter

$$(M + \mu\Lambda)^\cdot = [M + \mu\Lambda, \Omega + \mu\Lambda^2], \quad \Omega = \Lambda M + M\Lambda. \qquad\qquad (5.5)$$

The coefficients of μ in $Tr(M + \mu\Lambda)^k$ again yield integrals and it would be interesting to know if the singular generalized rigid body problems discussed here are integrable. We saw earlier this is true for $SO(3)$.

Another class of singular rigid body equations related to symmetric spaces is given by Fomenko [1982]. Recall firstly that for $X \in \mathfrak{g}$, $\mathfrak{g}$ a semisimple Lie algebra, Fomenko defines a class of generalized rigid body flows $\dot{X} = [X, \varphi_{a,b}X]$ where $\varphi_{a,b} = ad_a^{-1}ad_b$, $a, b \in \mathfrak{g}$ two commuting elements in general position.

Now consider a symmetric space G/K as defined above with the decomposition $\mathfrak{g} = \mathfrak{p} \oplus \mathfrak{k}$. One takes $a \in \mathfrak{p}$, $b \in \mathfrak{h} \subset \mathfrak{k}$ where $\mathfrak{h}$ is the annihilator of a, $\mathfrak{h} = \{h \in \mathfrak{k} \mid [h, a] = 0\}$.

We now decompose $\mathfrak{g}$ still further. Let $\mathfrak{h}'$ be an algebraic complement to $\mathfrak{h}$ in $\mathfrak{k}$, $\mathfrak{k} = \mathfrak{h} \oplus \mathfrak{h}'$, $\mathfrak{h} \cap \mathfrak{h}' = 0$. Let $\mathrm{Ker}\,(ad_b)$ be denoted M and note that $a \in M$. Define $B = M \cap \Phi_a\mathfrak{h}'$, $\Phi_a : \mathfrak{h} \to \mathfrak{p}$ given by $\Phi_a(k) = [k, a]$. Now let T be an algebraic complement to B in M, i.e. $M = T \oplus B$.

Fomenko considers for example the case $S^{n-1} = SO(n)/SO(n-1)$. Let $(ij) = E_{ij} - E_{ji}$, E_{ij} the matrix in $so(n)$ with 1 in the ijth position and zero elsewhere. $(1k)$, $2 \le k \le n$ is a basis for $\mathfrak{p}$ and an arbitrary element of $\mathfrak{p}$ may thus be written

$$P = \sum_{k=2}^{n} x_k(1k).$$

$\mathfrak{h}$ here is $so(n-2)$ embedded in the standard fashion and T is the span of (12). Fomenko thus takes b to be an arbitrary element of $so(n-2)$ and a to be (12).

The flow is not defined for $n \le 3$. The structure above enables one to restrict the flow to $\mathfrak{p}$.

We can compute the flow for $SO(4)/SO(3)$. Here an element of $\mathfrak{p}$ is of the form

$$P = \begin{bmatrix} 0 & x_1 & x_2 & x_3 \\ -x_1 & 0 & 0 & 0 \\ -x_2 & 0 & 0 & 0 \\ -x_3 & 0 & 0 & 0 \end{bmatrix}, \tag{5.6}$$

$a = (12)$ and $b \equiv b(34)$, b a scalar.

Then

$$ad_a^{-1} ad_b P = \begin{bmatrix} 0 & 0 & 0 & 0 \\ 0 & 0 & x_3 b & -x_2 b \\ 0 & -x_3 b & 0 & 0 \\ 0 & x_2 b & 0 & 0 \end{bmatrix}. \tag{5.7}$$

Hence

$$[P, ad_a^{-1} ad_b P] = \begin{bmatrix} 0 & 0 & x_1 x_3 b & -x_1 x_2 b \\ 0 & 0 & 0 & 0 \\ -x_1 x_3 b & 0 & 0 & 0 \\ x_1 x_2 b & 0 & 0 & 0 \end{bmatrix}. \tag{5.8}$$

Thus

$$\begin{aligned} \dot{x}_1 &= 0 \\ \dot{x}_1 &= x_1 x_3 b \\ \dot{x}_3 &= -x_1 x_2 b. \end{aligned} \tag{5.9}$$

We can set $x_1 = 1$ for example. Notice then that the flow in x_2 and x_3 is a rotation. $\square$

Here we have merely contrasted the Fomenko singular systems to our sub-Riemannian case. We hope to return to these systems in a future paper.

Acknowledgement: We would like to thank Roger Brockett for pointing out a number of interesting connections with his work in this area. We also wish to thank J. Baillieul and R. Montgomery for their comments and suggestions.

References

Adler, M., and van Moerbeke, P. [1980] *Completely Integrable Systems, Euclidean Lie Algebras and Curves*, Adv. Math. **38**, 267–317.

Arnold, V.I. [1978] *Mathematical Methods of Classical Mechanics*, Springer-Verlag, New York.

Baillieul, J. [1975] *Some Optimization Problems in Geometric Control Theory*, Ph.D. Thesis, Harvard University.

—— [1978] *Geometric methods for nonlinear optimal control problems*, J. of Optimization Theory and Applications, **25 No. 6**, 519–548.

Bloch, A.M., and Crouch, P.E. [1992a] *On the dynamics and control of nonholonomic systems on Riemannian manifolds*, Proc. NOLCOS '92, Bordeaux.

—— [1992b] *Nonholonomic and vakonomic control systems on Riemannian manifolds*, (to appear).

Brockett, R.W. [1973] *Lie theory and control systems defined on spheres*, SIAM J. Appl. Math., **23 No. 2**, 213–225.

______ [1982] *Control theory and singular Riemannian geometry*, in New Directions in Applied Mathematics (eds. P.J. Hilton and G.S. Young), Springer Verlag, New York.

Brockett, R.W., and Dai, Liyi [1992] *Nonholonomic kinematics and the role of elliptic functions in constructive controllability*, (to appear).

Bryant, R.L., and Hsu, L. [1993] *Rigidity of integral curves of rank 2 distributions*, **1993 No. 5**, Duke University Preprint.

Crouch, P.E. and Leite, F. Silva [1991a] *Geometry and the dynamic interpolation problem*, Proc. A.C.C., Boston, 1131–1136.

______ [1991b] *The dynamic interpolation problem on Riemannian manifolds, Lie Groups, and Symmetric Spaces*, (to appear).

Faybusovich, L.E. [1988] *Explicitly solvable nonlinear optimal control problems*, Int. J. Control, **48 No. 6**, 2507–2526.

Fomenko, A.T. [1982] *On symplectic structures and integrable systems on symmetric spaces*, Math. USSR Sbornik, **43, No. 2**, 235–250.

Fomenko, A.T., and Trofimov, V.V. [1988] *Integrable Systems on Lie Algebras and Symmetric Spaces*, Gordon and Breach, New York.

Griffiths, P.A. [1983] *Exterior Differential Systems*, Birkhauser, Boston.

Günther, N.C. [1982] *Hamiltonian Mechanics and Optimal Control*, Thesis, Harvard University.

Hermann, R. [1977] *Differential Geometry and the Calculus of Variations*, Interdisciplinary Mathematics, Vol. XVII, Math. Sci. Press, Brookline, Mass.

Jurdjevic, V. [1991a] *Nonlinear elastica*, (to appear).

______ [1991b] *Elastic rods, Radon's problem and the variational problems on Lie groups and their homogeneous spaces*, (to appear).

______ [1991c] *Optimal control problems on Lie groups*, Analysis of Controlled Dynamical Systems, Birkhaüser, Boston.

Krishnaprasad, P.S., and Yang, R. [1991] *Geometric phases, anholonomy and optimal movement*, Proc. of the Conf. on Robotics and Automation 1991, Sacramento, 2185–2189.

Lafferiere, G., and Sussmann, H.J. [1991] *Motion planning for completely nonholonomic systems without drift*, Proc. of the Conf. on Robotics and Automation, Sacramento, 1148–1153.

Manakov, S. [1976] *Note on the integration of Euler's equations of the dynamics of an n-dimensional rigid body*, Funct. Anal. Appl., **10**, 328–329.

Marsden, J.E., and Scheurle, J. [1992] *Lagrangian reduction and the double spherical pendulum*, (to appear).

Mishchenko, A.S., and Fomenko, A.T. [1976] *On the integration of Euler's equations on semisimple Lie algebras*, Dokl. Akad. Nauk. SSSR, **231 (3)**, 536–539.

______ [1978] *Euler equations on finite dimensional Lie groups*, Izv. Akad. Nauk. SSSR (Math Ser), **42 (2)**, 396–415.

______ [1982] *Integrability of Euler equations on semisimple Lie algebras*, Sel. Math. Sov., **2, No. 3**, 207–292.

Milnor, J. [1963] *Morse Theory*, Princeton University Press.

Montgomery, R. [1990] *Isoholonomic problems and some applications*, Comm. Math. Phys., **128**, 565–592.

——— [1993] *Geodesics which do not satisfy the geodesic equations*, (to appear).

Murray, R., and Sastry, S. [1990] *Steering nonholonomic systems using sinusoids*, Proc. of the 29th IEEE Conf. on Decision and Control, IEEE, New York, 2097–2101.

Nomizu, Y. [1954] *Invariant affine connections on homogeneous spaces*, Amer. J. Math., **6**, 33–65.

Ratiu, T. [1980] *The motion of the free n-dimensional rigid body*, Indiana U. Math Journal, **29**, 609–627.

Strichartz, R.S. [1986] *Sub-Riemannian geometry*, J. Diff. Geom., **24**, 221–263.

——— [1987] *Realms of mathematics: elliptic, hyperbolic, parabolic, sub-elliptic*, Math Intelligencer, **9 (no.3)**, 56–64.

Sussmann, H.J. [1992] *A cornucopia of abnormal sub-Riemannian minimizers*, (to appear).

Vershik, A.M. *Classical and nonclassical dynamics with constraints, in New in Global Analysis*, Voronteh, Gos. Univ.; English Trans. Lecture Notes in Math **1108** (1984).

Fields Institute Communications
Volume **3**, 1994

Systems of Hydrodynamic Type, Connected with the Toda Lattice and the Volterra Model

O.I. Bogoyavlenskij
Department of Mathematics
Queen's University
Kingston, Ontario, Canada
K7L 3N6

Abstract. A method for constructing systems of hydrodynamic type having a number of Riemann invariants is proposed. The systems of hydrodynamic type naturally connected with the Volterra model and Toda lattice are constructed together with their integrable extensions. Continuous limits of these systems are found, one of them an equation describing interaction between Riemann breaking waves and transversal KdV long waves.

1 Method for constructing nonlinear equations possessing a countable family of conservation laws

I. A certain class of integrable $(n + 1)$-dimensional equations was studied by Calogero and Degasperis [1976, 1977, 1980] by using the generalized Wronskian relations. A general Lax type operator equation was proposed by Zakharov [1983] for constructing $(n+1)$-dimensional integrable equations. These constructions were also discussed in the monograph by Dodd, Ellbeck, Gibbon and Morris [1982].

In the present work differential equations are studied which are equivalent to the following equation in space of linear operators L and A:

$$L_t = P(L) + \sum_{k=1}^{n} R_k(L, L_{y_k}) + [L, A], \qquad (1.1)$$

where $P(L)$ and $R_k(L, L_{y_k})$ are certain meromorphic functions of an operator L and the functions $R_k(L, L_{y_k})$ are linear with respect to L_{y_k}. We assume that the operators L and A depend on the variables $t, y_1, \ldots, y_n$ and $L_{y_k} = \partial L / \partial y_k$. It is supposed for concreteness that L and A are arbitrary finite-dimensional operators. But all results of Section 1 are valid as well L and A, which are differential operators in the variable x, if L is symmetric and A is skew-symmetric.

1991 *Mathematics Subject Classification.* Primary 35Q58; Secondary 35L55.

Supported in part by the Ministry of Colleges and Universities of Ontario and the Natural Sciences and Engineering Research Council of Canada while visiting The Fields Institute.

Coefficients of the meromorphic functions $P(L)$, $R_k(L, L_{y_k})$ are assumed to depend on invariants of the operator L and their derivatives with respect to variables $t, y_1, \ldots, y_n$, that is, the coefficients do not change after the transformation $L \to QLQ^{-1}$.

Lemma 1.1 *In view of equation (1.1) the eigenvalues $f(t, y_1, \ldots, y_n)$ of the operator L satisfy the equation*

$$f_t = P(f) + \sum_{k=1}^{n} R_k(f, f_{y_k}). \tag{1.2}$$

Proof Let us represent the operator L in the form

$$L = QL_0Q^{-1}, \tag{1.3}$$

where Q is a certain invertible operator. The following equalities hold

$$L_t = QL_{0t}Q^{-1} + Q_tQ^{-1}L - LQ_tQ^{-1}, \tag{1.4}$$

$$L_{y_k} = QL_{0y_k}Q^{-1} + Q_{y_k}Q^{-1}L - LQ_{y_k}Q^{-1}. \tag{1.5}$$

In view of the definitions of the functions $P(L)$ and $R_k(L, L_{y_k})$ we get

$$P(L) = QP(L_0)Q^{-1}, \tag{1.6}$$

$$R_k(L, L_{y_k}) =$$
$$QR_k(L_0, L_{0y_k})Q^{-1} + R_k(L, Q_{y_k}Q^{-1})L - LR_k(L, Q_{y_k}Q^{-1}). \tag{1.7}$$

After the substitution of the equalities (1.4)-(1.7) equation (1.1) assumes the form

$$Q(L_{0t} - P(L_0) - \sum_{k=1}^{n} R_k(L_0, L_{0y_k}))Q^{-1} =$$

$$[L, \quad Q_tQ^{-1} - \sum_{k=1}^{n} R_k(QL_0Q^{-1}, Q_{y_k}Q^{-1}) + A]. \tag{1.8}$$

This equation is valid if the following two equations hold:

$$L_{0t} = P(L_0) + \sum_{k=1}^{n} R_k(L_0, L_{0y_k}), \tag{1.9}$$

$$Q_t = -AQ + \sum_{k=1}^{n} R_k(QL_0Q^{-1}, Q_{y_k}Q^{-1})Q. \tag{1.10}$$

Equation (1.9) determines completely the evolution of the operator L_0. Equation (1.10) determines the evolution of Q.

Let us choose the operator $Q(t_0, y_1, \ldots, y_n)$ at an initial moment of time $t = t_0$ in such a way that the operator $L_0(t, y_1, \ldots, y_n)$ is reduced to its Jordan normal form (upper triangular). Then the right-hand side of equation (1.9) is also upper-triangular. Therefore the solution of equation (1.9), namely the operator $L_0(t, y_1, \ldots, y_n)$ with the indicated initial data, will be upper-triangular always. Hence the diagonal entries of the operator L_0 which are its eigenvalues, coinciding with the eigenvalues $f(t, y_1, \ldots, y_n)$ of the operator L (1.3), satisfy the equation (1.2). Lemma 1.1 is proved. $\square$

Remark 1.2 *If coefficients of the functions $P(L)$, $R_k(L, L_{y_k})$ are constants, then (1.2) is split into equations for each eigenvalue $f(t, y_1, \ldots, y_n)$ separately. If these coefficients depend on invariants of the operator L, then (1.2) is a system of equations connecting the eigenvalues. In the case when coefficients of the functions $P(L)$, $R_k(L, L_{y_k})$ depend on derivatives of invariants of the operator L with respect to variables $t, y_1, \ldots, y_n$ we obtain the system of equations (1.2) with derivatives of an arbitrary order.*

II. Let the vector $\phi(t, y_1, \ldots, y_n)$ be the eigenfunction of the operator L, corresponding to the eigenvalue $f(t, y_1, \ldots, y_n)$:

$$(L - f)\phi = 0. \tag{1.11}$$

In view of linearity with respect to operator L_{y_k}, the function $R_k(L, L_{y_k})$ may be presented in the form

$$R_k(L, L_{y_k}) = \sum_m S_{km}(L) L_{y_k} T_{km}(L). \tag{1.12}$$

Lemma 1.3 *Eigenfunctions ϕ of the operator L, in view of the equation (1.1), satisfy the equation*

$$(L - f)(\phi_t + A\phi - \sum_{k=1}^{n} \sum_m S_{km}(L) T_{km}(f) \phi_{y_k}) = 0. \tag{1.13}$$

Proof Differentiating the equality (1.11) with respect to time t and y_k, we obtain

$$(L_t - f_t)\phi + (L - f)\phi_t = 0, \tag{1.14}$$

$$(L_{y_k} - f_{y_k})\phi + (L - f)\phi_{y_k} = 0. \tag{1.15}$$

Equation (1.14) after the substitution of the expression for L_t (1.1) assumes the form

$$(P(L) + \sum_{k=1}^{n} \sum_m S_{km}(L) L_{y_k} T_{km}(L) + LA - AL - f_t)\phi + (L - f)\phi_t = 0. \tag{1.16}$$

Hence, using the equations (1.14) and (1.15) and in view of the commutativity of the functions of the operator L we get the equation

$$(L - f)(\phi_t + A\phi - \sum_{k=1}^{n} \sum_m S_{km}(L) T_{km}(f) \phi_{y_k})$$

$$+ (P(f) + \sum_{k=1}^{n} \sum_m S_{km}(f) T_{km}(f) f_{y_k} - f_t)\phi = 0. \tag{1.17}$$

According to Lemma 1.1, the eigenvalue $f(t, y_1, \ldots, y_n)$ satisfies the equation (1.2). Therefore the equation (1.13) follows from the equations (1.17) and (1.2). Lemma 1.3 is proved. $\square$

Remark 1.4 *If the spectrum of the operator L is nondegenerate then the eigenfunctions $\phi(t, y_1, \ldots, y_n)$ in view of equation (1.13), may be normalized so that the following equation holds:*

$$\phi_t + A\phi - \sum_{k=1}^{n} \sum_{m} S_{km}(L) T_{km}(f) \phi_{y_k} = 0. \tag{1.18}$$

Let the function $P(f) = 0$. Then equation (1.18), in view of (1.2), is invariant under transformations

$$\phi \to F(f)\phi, \tag{1.19}$$

where $F(f)$ is an arbitrary function.

Remark 1.5 *Let the operator L be symmetric, and let A be skew-symmetric. Then equation (1.18) leads to the equation for the scalar product (ϕ, ϕ):*

$$(\phi, \phi)_t - \sum_{k=1}^{n} \sum_{m} S_{km}(f) T_{km}(f)(\phi, \phi)_{y_k} = 0. \tag{1.20}$$

This equation coincides with the equation (1.2) as $P(f) = 0$. Hence as $n = 1$ it follows in view of the gauge-invariance (1.19) of the equation (1.18) that the eigenfunction $\phi(t, y)$ may be normalized by a transformation (1.19) in such a way that $(\phi, \phi) = 1$ on every segment where $f_y \neq 0$.

III. Let us consider the following special case of equation (1.1):

$$L_t = \sum_{k=1}^{n} L_{y_k} T_k(L). \tag{1.21}$$

This equation, being written for entries of the matrix L is a system of hydrodynamic type in the terminology of the work in Whitham [1965], and Dubrovin and Novikov [1989].

Proposition 1.6 *If the spectrum of the operator L is nondegenerate, then the equation (1.21) admits a representation in the Riemann invariants, which are the eigenvalues and the eigenfunctions of the operator L.*

Proof According to Lemma 1.1 the following equations are valid for the eigenvalues $f(t, y_1, \ldots, y_n)$ of the operator L:

$$f_t = \sum_{k=1}^{n} T_k(f) f_{y_k}. \tag{1.22}$$

In view of Lemma 1.3 we get the equations for the eigenfunctions $\phi(t, y_1, \ldots, y_n)$:

$$(L - f)(\phi_t - \sum_{k=1}^{n} T_k(f) \phi_{y_k}) = 0. \tag{1.23}$$

Hence by virtue of the nondegeneracy of the spectrum of the operator L it follows the eigenfunction $\phi(t, y_1, \ldots, y_n)$ may be normalized in such a way that the equation holds

$$\phi_t = \sum_{k=1}^{n} T_k(f) \phi_{y_k}. \tag{1.24}$$

In view of the equations (1.22) and (1.24) the eigenvalues $f(t, y_1, \ldots, y_n)$ and the eigenfunctions $\phi(t, y_1, \ldots, y_n)$ are the Riemann invariants. It is obvious that their evolution determines completely the evolution of the operator L. $\square$

IV. Let us consider the equation (1.1) under the following assumptions:

$$P(L) \equiv 0, \quad R_k(L, L_{y_k}) = \sum_{0 \le i \le m}^{\infty} c_{ik}^m L^{m-i} L_{y_k} L^i, \tag{1.25}$$

where coefficients c_{ik}^m are constant, $0 \le m < \infty$ and series $|c_{ik}^m|$ is convergent.

Lemma 1.7 *Equation (1.1) under the assumptions (1.25) has an infinite set of conservation laws.*

Proof When coefficients of functions $P(L)$ and $R_k(L, L_{y_k})$ are constants, the equation (1.2) is split into a system of noninteracting equations for each eigenvalue $f(t, y_1, \ldots, y_n)$ separately. We get the following consequence from the equation (1.2) under assumptions (1.25),

$$(f^p)_t = \sum_{k=1}^{n} \left(\sum_{0 \le i \le m}^{\infty} \frac{p c_{ik}^m}{m+p} f^{m+p} \right)_{y_k}, \tag{1.26}$$

where p is an arbitrary positive number. Hence, assuming that the eigenvalues $|f(t, y_1, \ldots, y_n)|$ tend to zero rapidly enough for $|y_k| \to \infty$ and applying Gauss's theorem, we obtain the conserved quantities

$$\frac{dJ_p}{dt} = 0, \quad J_p = \int_{R^n} f^p dy_1 \ldots dy_n. \tag{1.27}$$

Thus the functionals J_p (1.27) form an infinite set of conserved quantities for the equation (1.1) under assumptions (1.25).

In the case of finite-dimensional operators L, other conservation laws are more convenient. Let us multiply the equation (1.1) by pL^{p-1} where p now is an arbitrary integer, $p \ge 1$. After taking trace of matrices we obtain the equation

$$(\mathrm{Tr} L^p)_t = \sum_{k=1}^{n} \left(\sum_{0 \le i \le m}^{\infty} \frac{p c_{ik}^m}{m+p} \mathrm{Tr} L^{m+p} \right)_{y_k}. \tag{1.28}$$

Equation (1.28) for all natural integers p is a conservation law. If traces of degrees of the matrix L tend to zero rapidly enough as $|y_k| \to \infty$ then, applying Gauss's theorem to the equation (1.28), we get the equation

$$\frac{dI_p}{dt} = 0, \quad I_p = \int_{R^n} \mathrm{Tr}(L^p) dy_1 \ldots dy_n. \tag{1.29}$$

Thus, the functionals I_p (1.29) form a countable set of conserved quantities for the equation (1.1) under assumptions (1.25). Lemma 1.7 is proved. $\square$

In the solutions of the equations (1.1) under assumptions (1.25), there occurs a breaking (overlapping) of the wave front $f(t, y_1, \dots, y_n)$ (eigenvalue of the operator L), this is valid for all variable solutions for $n = 1$. Therefore the phenomenon of breaking of some matrix entry $L_{ik}(t, y_1, \dots, y_n)$ in the case of finite-dimensional operators L is a notable feature of the general solution of equation (1.1).

All the properties indicated in this section distinguish essentially the equation (1.1) from the Lax equation $\dot{L} = [L, A]$. The equation (1.1) as $P(L) \neq 0$, $R_k(L, L_{y_k}) = 0$ has attractors in the phase space, see Bogoyavlenskij [1987].

Remark 1.8 *Equation (1.1), with $P(L) = 0$, defines the hierarchy of equations with breaking solutions, corresponding to the operator L. It is important to note that, in contrast to the Lax equation, where the order of the main derivative with respect to x (in the differential operator A) is the parameter of hierarchy, for the operator equation (1.1), the hierarchy becomes $(n+1)$-dimensional and has an additional parameter which is the number of new independent variables $y_1, \dots, y_n$. Particularly, the simplest matrix operators (2.1) and (4.1) considered below generate the hierarchies of $(n+1)$-dimensional equations, admitting the operator representation (1.1).*

2 System of hydrodynamic type, connected with the Volterra model

I. Let L be a symmetric matrix, having the Jacobi form

$$
\begin{pmatrix}
0 & a_1 & 0 & \dots & 0 \\
a_1 & 0 & a_2 & \dots & 0 \\
0 & a_2 & 0 & \dots & 0 \\
\vdots & \vdots & \vdots & \vdots & \vdots \\
0 & \dots & a_{n-2} & 0 & a_{n-1} \\
0 & 0 & \dots & a_{n-1} & 0
\end{pmatrix}.
\tag{2.1}
$$

The matrix A is skew-symmetric and has only the following non-zero entries

$$
A_{i,i+2} = x_i, \qquad A_{i+2,i} = -x_i, \qquad i = 1, \dots, n-2.
\tag{2.2}
$$

Let us consider for these matrices L and A the operator equation

$$
L_t = LL_y L + [L, A].
\tag{2.3}
$$

This equation is equivalent to the system of equations

$$
a_i(a_{i+1})_y a_{i+2} = a_{i+2} x_i - a_i x_{i+1}, \qquad i = 1, \dots, n-3,
\tag{2.4}
$$

$$
(a_i)_t = \frac{1}{2} a_i((a_{i-1}^2)_y + (a_i^2)_y + (a_{i+1}^2)_y) + a_{i-1} x_{i-1} - a_{i+1} x_i.
\tag{2.5}
$$

After the substitution

$$
x_i = a_i a_{i+1} b_i,
\tag{2.6}
$$

equation (2.4) assumes the form

$$
(a_{i+1})_y a_{i+1}^{-1} = b_i - b_{i+1}.
\tag{2.7}
$$

Solution of the system (2.7) is given by the formulae (β is an arbitrary function of t, y)

$$
b_i = -\sum_{k=1}^{k=i} (a_k)_y a_k^{-1} - \beta/2.
\tag{2.8}
$$

System (2.5), after the substitution of the formulae (2.6) and (2.8) takes the form

$$(a_i)_t = \frac{1}{2}a_i\Big((a_i^2)_y + 2a_{i+1}^2 \sum_{k=1}^{k=i+1}(a_k)_y a_k^{-1} - 2a_{i-1}^2 \sum_{k=1}^{k=i-2}(a_k)_y a_k^{-1}$$
$$+\beta(a_{i+1}^2 - a_{i-1}^2)\Big). \tag{2.9}$$

Thus, the system (2.9) admits the operator representation (2.3) and therefore, according to Lemma 1.1, the eigenvalues $f_k(t,y)$ of the corresponding matrix L satisfy the equations

$$f_{k_t} = f_k^2 f_{k_y}. \tag{2.10}$$

The system (2.9) as $\beta = 0$ belongs to the class of hydrodynamic type systems (Whitham [1965], Dubrovin and Novikov [1989]). The eigenvalues $f_k(t,y)$ in view of equations (2.10) are the Riemann invariants. The number of independent Riemann invariants is equal to $[n/2]$, because the eigenvalues of Jacobi matrix (2.1) are symmetric with respect to zero.

System (2.9) for the solutions, independent of y is transformed into the classical Volterra system. System (2.9) as $\beta = m/y$ admits the following reduction

$$a_i^2(t,y) = yc_i(t), \tag{2.11}$$

and is reduced to the dynamical system

$$\dot{c}_i = c_i((m+i)(c_{i+1} - c_{i-1}) + 2c_{i+1} + c_i + c_{i+1}), \tag{2.12}$$

derived in Bogoyavlenskij [1990]. System (2.12) is equvalent to the operator equation

$$L_t = L^3 + [L, A], \tag{2.13}$$

where matrices L and A have the same form (2.1) and (2.2).

After the substitution of $a_i = \exp(u_i/2)$ the system (2.9) assumes the form

$$u_{i_t} = e^{u_i}u_{i_y} + e^{u_{i+1}}\sum_{k=1}^{k=i+1}u_{k_y} - e^{u_{i-1}}\sum_{k=1}^{k=i-2}u_{k_y} + \beta(e^{u_{i+1}} - e^{u_{i-1}}). \tag{2.14}$$

II. Let $\phi(t,y)$ be the eigenfunction of the matrix L, corresponding to the eigenvalue $f(t,y)$. Denote by $\phi_k(t,y)$ its coordinates: $\phi_k(t,y) = (\phi, e_k)$, where vectors $e_1, \dots, e_n$ form a basis. In view of (2.1) we obtain from (1.11) the equations $a_1\phi_2 = f\phi_1$, $a_1\phi_1 + a_2\phi_3 = f\phi_2$. Hence the following equalities,

$$\phi_2 = fa_1^{-1}\phi_1, \qquad \phi_3 = (f^2 - a_1^2)a_1^{-1}a_2^{-1}\phi_1. \tag{2.15}$$

The spectrum of the general Jacobi matrix (2.1) is nondegenerate. Therefore it follows from Lemma 1.3 that the eigenfunction $\phi(t,y)$ may be normalized so that the equation holds

$$\phi_t + A\phi - fL\phi_y = 0. \tag{2.16}$$

Projecting this equation onto vector e_1 we get

$$\phi_{1_t} + x_1\phi_3 - fa_1\phi_{2_y} = 0. \tag{2.17}$$

In view of formulae (2.6) and (2.8) for $i = 1$ we have

$$x_1 = a_1a_2(-a_{1_y}a_1^{-1} - \beta/2). \tag{2.18}$$

Equation (2.17) after the substitution of the formulae (2.15) and (2.18) assumes the form

$$\phi_{1_t} - f^2 \phi_{1_y} = \frac{1}{2}((f^2 - a_1^2)_y + \beta(f^2 - a_1^2))\phi_1. \tag{2.19}$$

The eigenfunctions $\phi(t, y)$ of the symmetric operator L are orthogonal. According to Remark 1.5, they may be normalized in such a way that the equation (2.16) is valid. Thus we get

$$e_1 = \sum_{k=1}^{n} \phi_1(f_k)\phi(f_k) \tag{2.20}$$

and hence

$$a_1^2 = \left(L^2 \sum_{k=1}^{n} \phi_1(f_k)\phi(f_k), e_1\right) = \sum_{k=1}^{n} f_k^2(\phi_1(f_k))^2. \tag{2.21}$$

In view of the orthonormality of the eigenfunctions we have

$$\sum_{k=1}^{n} (\phi_1(f_k))^2 = 1. \tag{2.22}$$

Therefore, functions $r_k(t, y)$ exist for which

$$\phi_1(f_k) = r_k \left(\sum_{s=1}^{n} r_s^2\right)^{-1/2}. \tag{2.23}$$

Let us consider the system of equations

$$r_{k_t} - f_k^2 r_{k_y} = \frac{1}{2}\left((f_k^2)_y - f_k^2\left(\ln \sum_{s=1}^{n} r_s^2\right)_y + \beta f_k^2\right)r_k. \tag{2.24}$$

It is simple to check that the system of equations (2.19) and (2.21) follow from the system (2.24) after the transformation (2.23).

Let us introduce, as in Moser [1975], the resolvent operator $R(\lambda) = (\lambda - L)^{-1}$ and the function

$$f(\lambda) = (R(\lambda)e_1, e_1) = \sum_{k=1}^{n} \frac{(\phi_1(f_k))^2}{\lambda - f_k}. \tag{2.25}$$

The coefficients $\phi_1(f_k)$ coincide, corresponding to pairs of eigenvalues f_k symmetric with respect to zero. The entries $a_1, \dots, a_{n-1}$ of the matrix L are functions of the coefficients r_k (2.23). Therefore the system of equations (2.10) and (2.24) are equivalent to the system of equations in (2.13). For the solutions independent of y systems (2.10) and (2.24) are reduced to the simplest system of equations

$$f_{k_t} = 0, \quad r_{k_t} = \frac{1}{2}\beta f_k^2 r_k = \frac{\partial}{\partial r_k}\left(\frac{\beta}{4}\sum_{s=1}^{n} f_s^2 r_s^2\right), \tag{2.26}$$

having gradient form, which were derived for the Volterra model in Moser [1975].

Equations (2.10) are represented in the form, similar to the gradient one

$$f_{k_t} = \frac{\partial}{\partial y}\frac{\partial V_0}{\partial f_k}, \quad V_0 = \frac{1}{12}Tr(L^4).$$

Equations (2.24) in variables $\alpha_k = r_k^2$ have the form

$$\alpha_{k_t} = a\frac{\partial}{\partial y}\left(\frac{1}{a}\frac{\partial V_1}{\partial \alpha_k}\right) + \beta\frac{\partial V_1}{\partial \alpha_k}, \quad a = \sum_{s=1}^{n}\alpha_s, \quad V_1 = \frac{1}{2}\sum_{s=1}^{n}f_s^2\alpha_s^2.$$

Hence in variables $\psi_n = (\phi_1(f_k))^2 = \alpha_k a^{-1}$ we obtain the system

$$\psi_{k_t} = \frac{\partial}{\partial y}\frac{\partial V_2}{\partial \psi_k} - \psi_k\frac{\partial}{\partial y}\sum_{s=1}^{n}\frac{\partial V_2}{\partial \psi_s} + \beta\frac{\partial V_2}{\partial \psi_k} - \beta\psi_k\sum_{s=1}^{n}\frac{\partial V_2}{\partial \psi_s}$$

$$V_2 = \frac{1}{2}\sum_{s=1}^{n}f_s^2\psi_s^2.$$

Remark 2.1 *Equation (2.3) is equivalent to the Lax equation of the following form:*

$$L_t = [L,\, A - L\partial_y L]. \tag{2.27}$$

Therefore the constructed systems of equations (2.9) and (2.14) admit also the Lax representation (2.27).

III. System (2.14) has the following form

$$u_{i_t} = \sum_{j=1}^{n-1}A^{ij}\frac{\partial H}{\partial u_j}, \tag{2.28}$$

$$H = \frac{1}{2}\mathrm{Tr}L^2 = e^{u_1} + e^{u_2} + \cdots + e^{u_{n-1}}. \tag{2.29}$$

The operators A^{ij} are skew-symmetric ones and have the form

$$A^{ij} = g^{ij}\frac{\partial}{\partial y} + \sum_{k=1}^{n-1}b_k^{ij}u_{k_y} + \beta I^{ij}. \tag{2.30}$$

Here the coefficients g^{ij} are symmetric, I^{ij} and b_k^{ij} are skew-symmetric. They all are constant and non-zero only in the following cases

$$g^{ii} = g^{i,i+1} = g^{i+1,i} = 1, \qquad I^{i,i+1} = -I^{i+1,i} = 1, \tag{2.31}$$

$$b_k^{i,i+1} = -b_k^{i+1,i} = 1 \qquad \text{as } 1 \le k \le i.$$

The operators A^{ij} (2.30) have the form of ones defining the Poisson brackets of hydrodynamic type (Dubrovin and Novikov [1989]). The equality

$$\partial(g^{ij})/\partial u^k = b_k^{ij} + b_k^{ji}, \tag{2.32}$$

holds, in view of which the Christoffel symbols

$$\Gamma_{jk}^{i} = -g_{jm}b_k^{mi} \tag{2.33}$$

determine the constant differential-geometric connection compatible with the metric g_{ij}.

In view of the equality (2.32) the scalar product of two functionals $F(u)$ and $G(u)$

$$\langle F, G\rangle = \int_{-\infty}^{\infty}\frac{\delta F}{\delta u_i}A^{ij}\frac{\delta G}{\delta u_j}\,dy, \tag{2.34}$$

is skew-symmetric.

The coefficients g^{ij}, b_k^{ij} are represented in the form

$$g^{ij} = \gamma^{ij} + \gamma^{ji}, \qquad b_k^{ij} = \frac{\partial \gamma^{ij}}{\partial u^k}. \tag{2.35}$$

Here the coefficients γ^{ij} are non-zero only in the following cases:

$$\gamma^{ii} = \frac{1}{2}, \quad \gamma^{i,i+1} = \frac{1}{2} + S_i, \quad \gamma^{i+1,i} = \frac{1}{2} - S_i, \quad S_i = \sum_{k=1}^{k=i} u_k.$$

In view of (2.35), and according to the terminology of the work in Dubrovin and Novikov [1989], coordinates $u_1, \dots, u_{n-1}$ are called Liouville coordinates.

We note that the constant differential-geometric connection (2.33) in view of the formulae (2.31) is not a symmetric one and its torsion tensor T_{jk}^i and curvature tensor R_{jkl}^i are non-zero. Thus, in view of Theorem 1 of Dubrovin and Novikov [1989], the operator (2.30) does not satisfy the Jacobi identity for the Poisson brackets.

3 Continuous limits

I. System (2.9) assumes the following form after the substitution of $u_i = a_i^2$

$$u_{i_t} = u_i \Big(u_{i_y} + u_{i+1,y} +$$

$$u_{i+1} \sum_{k=1}^{k=i} u_{k_y} u_k^{-1} - u_{i-1} \sum_{k=1}^{k=i-2} u_{k_y} u_k^{-1} + \beta(u_{i+1} - u_{i-1}) \Big).$$

$$\tag{3.1}$$

We suppose that there exists a smooth function $v(t, x, y)$ such that

$$u_j = 1 - \varepsilon^2 v(t, x_j, y), \quad x_j = j\varepsilon, \quad \beta = 3\beta_0 \varepsilon^{-1}. \tag{3.2}$$

Substituting the expressions (3.2) into the equation (3.1), we have

$$v_{i_t} = \left(1 - \varepsilon^2 v_i\right)\left(2v_{i_y} + v_{i-1,y} + v_{i+1,y} - \varepsilon^2(v_{i+1} - v_{i-1}) \sum_{k=1}^{k=i} v_{k_y}\right.$$

$$\left. + \beta(v_{i+1} - v_{i-1}) + O(\varepsilon^3)\right), \tag{3.3}$$

where $v_i = v(t, x_i, y)$.

After applying the Taylor power series for the functions v and v_y and using the formula

$$\varepsilon \sum_{k=0}^{k=i} v_{k_y} = \int_0^{x_i} v_y(t, \xi, y) d\xi + O(\varepsilon) \tag{3.4}$$

the equation (3.3) is transformed into the equation

$$v_{i_t} = \left(1 - \varepsilon^2 v_i\right)\left(4v_{i_y} + \varepsilon^2\left(v_{i_{xxy}} - 2v_{i_x} \int_0^x v_y(t, \xi, y) d\xi\right.\right.$$

$$\left.\left. + 6\beta_0 v_{i_x} + \varepsilon 2\beta_0 v_{i_{xxx}} + O(\varepsilon^3)\right)\right). \tag{3.5}$$

Equation (3.5), after substituting:

$$t' = -\varepsilon^2 t, \quad y' = y + 4t, \quad x' = x + 6\beta_0 t \tag{3.6}$$

takes the form

$$v_{i_t} = 4v_i v_{i_y} + 2v_{i_x} \int_0^x v_y(t,\xi,y)d\xi - v_{i_{xxy}} + \beta_0(6v_i v_{i_x} - v_{i_{xxx}}) + O(\varepsilon). \qquad (3.7)$$

(The primes of the coordinates x', y', t' have been dropped.) The equation (3.7) is transformed in the limit $\varepsilon \to 0$ into the equation

$$v_t = 4vv_y + 2v_x \int_0^x v_y(t,\xi,y)d\xi - v_{xxy} + \beta_0(6vv_x - v_{xxx}). \qquad (3.8)$$

This equation belongs to the class of equations studied by Calogero and Degasperis [1976, 1977, 1980].

Equation (3.8) describes an interaction between Riemann breaking waves propagating along the y-axis and KdV long waves, propagating in transversal directions (Bogoyavlenskij [1989], [1990], [1991]). It follows from the results of that equation (3.8) has the following operator representation

$$L_t = 2(LL_y + L_yL) + [L, A], \qquad (3.9)$$

where $L = -\partial_x^2 + u(t,x,y)$ is the Sturm-Liouville operator. The detailed study of the equation (3.8) is carried out in Bogoyavlenskij [1990, 1991].

II. Let us indicate another continuous limit of the system (3.1). We assume a smooth function $v(t,x,y)$ exists satisfying the equalities

$$u_j(t,y) = v(t,x_j,y), \quad x_j = j\varepsilon, \quad \beta = \beta_0 \varepsilon^{-1}. \qquad (3.10)$$

After the substitution of (3.10), and with the limit $\varepsilon \to 0$, equation (3.1) is transformed into the equation

$$v_t = 4vv_y + 2vv_x \int_0^x v_y v^{-1}(t,\xi,y)d\xi + 2\beta_0 vv_x, \qquad (3.11)$$

which is the second continuous limit of the hydrodynamic type system (3.1).

After substituting $v = \exp u_x$, equation (3.11) takes the form

$$\frac{1}{2}u_{xt} = (e^{u_x}u_y)_x + (e^{u_x})_{xy} + \beta_0(e^{u_x})_x. \qquad (3.12)$$

III. In view of Lemma 1.7, the system of equations (3.1) has a countable set of conservation laws (1.28) and (1.29), which are non-trivial as $p = 2k$. The first integrals (1.29) in the continuous limit (3.11) assume the form

$$\frac{dI_k}{dt} = 0, \qquad I_k = \iint_{-\infty}^{\infty} v^k(t,x,y)dxdy. \qquad (3.13)$$

Let us derive the first integrals (3.13) for the equation (3.11) to (3.12) in an independent way. In view of (3.11), the following equation holds for an arbitrary real parameter k

$$\frac{k+1}{2k}(v^k)_t = \left(v^{k+1}\left(\beta_0 + \int_0^x v_y v^{-1}(t,\xi,y)d\xi\right)\right)_x + \frac{2k+1}{k+1}(v^{k+1})_y. \qquad (3.14)$$

Integrating this equation with respect to x and y, and assuming that $|v(t,x,y)| \to 0$ rapidly enough as $|x|, |y| \to \infty$, we get the first integrals (3.13) for an arbitrary $k > 0$.

The existence of the first integrals (3.13) obviously proves the presence of the solutions of equation (3.11) which are localized on the plane x, y for all moments of time.

We note that the same first integrals (3.13) exist as well for the whole hierarchy of $(n+1)$-dimensional equations (1.1) for $P(L) = 0$ with the Jacobi operator L (2.1).

4 Integrable extension of the hydrodynamic type system, connected with the Volterra model

In this paragraph we construct a new system of equations admitting the Lax representation and including as the special case system of hydrodynamic type (2.9). Let us consider the Lax equation of the form

$$(L + \alpha \partial_y)_t = \left[L + \alpha \partial_y, \frac{1}{3} \gamma \alpha^{-1} L^3 + A \right], \tag{4.1}$$

where matrices L and A have only the following non-zero entries

$$L_{i,i+1} = L_{i+1,i} = a_i, \tag{4.2}$$

$$A_{i,i+1} = A_{i+1,i} = z_i, \quad A_{i,i+2} = -A_{i+2,i} = x_i. \tag{4.3}$$

Equation (4.1) is equivalent to the matrix equation

$$L_t = \frac{1}{3} \gamma (L^3)_y + \alpha A_y + [L, A], \tag{4.4}$$

which after the substitution of the formulae (4.2) and (4.3) is reduced to the system of equations

$$a_{it} = \frac{1}{3} \gamma (a_i(a_{i-1}^2 + a_i^2 + a_{i+1}^2))_y + a_{i-1}x_{i-1} - a_{i+1}x_i + \alpha z_{iy}, \tag{4.5}$$

$$\frac{1}{3} \gamma (a_i a_{i+1} a_{i+2})_y = a_{i+2} x_i - a_i x_{i+1}, \tag{4.6}$$

$$\alpha x_{iy} = a_{i+1} z_i - a_i z_{i+1}. \tag{4.7}$$

After substituting

$$x_i = a_i a_{i+1} b_i, \qquad z_i = a_i c_i \tag{4.8}$$

into equations (4.6) and (4.7) we obtain the form

$$\frac{1}{3} \gamma \ln(a_i a_{i+1} a_{i+2})_y = b_i - b_{i+1}, \tag{4.9}$$

$$\alpha(b_i \ln(a_i a_{i+1})_y + b_{iy}) = c_i - c_{i+1}. \tag{4.10}$$

Solution of equations (4.9) is determined by the formulae

$$b_k = -\gamma \left(\sum_{j=1}^{k-1} \ln a_{j\,y} + \frac{2}{3} \ln a_{k\,y} + \frac{1}{3} \ln a_{k+1\,y} \right) - \frac{1}{2}\beta. \tag{4.11}$$

From equation (4.10) we find

$$c_k = -\alpha \Big(\sum_{j=1}^{k-1} b_j \ln(a_j a_{j+1})_y + \sum_{j=1}^{k-1} \ln b_{jy} + \delta \Big). \tag{4.12}$$

Equation (4.5), after the substitution of the formulae (4.8) and (4.11) and (4.12) assumes the form

$$a_{kt} = a_k \left(\frac{1}{2}\gamma(a_k^2)_y + \gamma a_{k+1}^2 \sum_{j=1}^{k+1} \ln a_{jy} - \gamma a_{k-1}^2 \sum_{j=1}^{k-2} \ln a_{jy} \right.$$

$$\left. + \frac{1}{2}\beta(a_{k+1}^2 - a_{k-1}^2) + \alpha(c_{ky} + c_k \ln a_{ky}) \right). \tag{4.13}$$

System (4.13) is the system of equations with third order derivatives with respect to variable y. When $\alpha = \beta = 0$ we get the hydrodynamic type system (2.9). Equalities (4.11) and (4.12) as $\gamma = 0$ lead to

$$b_k = -\frac{1}{2}\beta, \quad c_k = \alpha\beta \sum_{j=1}^{k-1} \ln a_{jy} + \frac{1}{2}\alpha\beta \ln a_{ky}. \tag{4.14}$$

In this case system (4.13) obtains the form

$$a_{kt} = a_k \left(\frac{1}{2}\beta(a_{k+1}^2 - a_{k-1}^2) + \alpha^2\beta \sum_{j=1}^{k-1} \ln a_{jyy} + \frac{1}{2}\alpha^2\beta \ln a_{kyy} \right.$$

$$\left. + \alpha^2\beta \ln a_{ky} \sum_{j=1}^{k-1} \ln a_{jy} + \frac{1}{2}\alpha^2\beta(\ln a_{ky})^2 \right). \tag{4.15}$$

After the substitution of $a_k = \exp u_k/2$ we get the system of equations

$$u_{kt} = \beta e^{u_{k+1}} - \beta e^{u_{k-1}} + \alpha^2\beta \sum_{j=1}^{k-1} u_{jyy} + \frac{1}{2}\alpha^2\beta u_{kyy}$$

$$+ \frac{1}{2}\alpha^2\beta u_{ky} \sum_{j=1}^{k-1} u_{jy} + \frac{1}{4}\alpha^2\beta(u_{ky})^2. \tag{4.16}$$

This system is also a two-dimensionalization of the Volterra system. Implicit form of such two-dimensionalization was mentioned in Mikhailov [1979].

System (4.15) and (4.16) in the continuous limit is transformed into the Kadomtzev-Petviashvili equation. The continuous limit of the system (4.13) is the equation with third order derivatives with respect to y, which generalizes the KP equation and equation of interaction of Riemann breaking waves with transversal KdV long waves.

System (4.13) has another equivalent Lax representation

$$(L + \alpha\partial_y)_t = \left[L + \alpha\partial_y, A - \frac{1}{3}\gamma A_1 \right],$$

$$A_1 = L^2\partial_y + L\partial_y L + \partial_y L^2 + \alpha(L\partial_y^2 + \partial_y L\partial_y + \partial_y^2 L) + \alpha^2\partial_y^3, \tag{4.17}$$

which is transformed in the continuous limit $\varepsilon \to 0$ into the Lax representation for the hydrodynamic type system (2.9).

5 System of hydrodynamic type, connected with the Toda lattice

I. Let L be a symmetric matrix of the form

$$
\begin{pmatrix}
p_1 & a_1 & 0 & \cdots & 0 \\
a_1 & p_2 & a_2 & \cdots & 0 \\
0 & a_2 & p_3 & \cdots & 0 \\
\vdots & \vdots & \vdots & \vdots & \vdots \\
0 & \cdots & a_{n-2} & p_{n-1} & a_{n-1} \\
0 & \cdots & 0 & a_{n-1} & p_n
\end{pmatrix}.
\tag{5.1}
$$

Let the matrix A be skew-symmetric and have only the following non-zero entries

$$
A_{i,i+1} = x_i, \qquad A_{i+1,i} = -x_i.
\tag{5.2}
$$

We consider for these matrices the operator equation

$$
L_t = LL_y + L_y L + [L, A].
\tag{5.3}
$$

Equation (5.3) is equivalent to the following system of equations

$$
(a_i a_{i+1})_y = x_i a_{i+1} - a_i x_{i+1},
\tag{5.4}
$$

$$
a_{i_t} = (a_i(p_i + p_{i+1}))_y + x_i(p_i - p_{i+1}),
\tag{5.5}
$$

$$
p_{i_t} = (p_i^2 + a_i^2 + a_{i-1}^2)_y + 2a_{i-1}x_{i-1} - 2a_i x_i.
\tag{5.6}
$$

After substitution of $x_i = a_i b_i$ the equations (5.4) assume the form

$$
a_{i_y} a_i^{-1} + a_{i+1,_y} a_{i+1}^{-1} = b_i - b_{i+1}.
\tag{5.7}
$$

The solution of this system of equations is given by the formulae

$$
-b_i = \beta + 2 \sum_{k=1}^{k=i-1} a_{k_y} a_k^{-1} + a_{i_y} a_i^{-1},
\tag{5.8}
$$

where β is an arbitrary function of t, y.

After the substitution of expression (5.8), and changing $\tilde{a}_i = a_i^2$, $\tilde{t} = 2t$ from equations (5.5) and (5.6), we have

$$
p_{i_t} = p_i p_{i_y} + a_i \sum_{k=0}^{k=i} a_{k_y} a_k^{-1} - a_{i-1} \sum_{k=1}^{k=i-2} a_{k_y} a_k^{-1} + \beta(a_i - a_{i-1}),
\tag{5.9}
$$

$$
a_{i_t} = p_{i+1} a_{i_y} + a_i(p_i + p_{i+1})_y + a_i(p_{i+1} - p_i)\Big(\beta + \sum_{k=1}^{k=i-1} a_{k_y} a_k^{-1}\Big),
$$

where we have omitted the tildes of $\tilde{a}_i$ and $\tilde{t}$.

The obtained system of equations (5.9) is equivalent to the operator equation (5.3). Therefore, according to Lemma 1.1, the eigenvalues $f_k(t, y)$ of the matrix L (5.1), due to the system (5.9), satisfy the equations

$$
f_{k_t} = 2f_k f_{k_y}.
\tag{5.10}
$$

Hence the eigenvalues $f_k(t, y)$ are the Riemann invariants. Thus system (5.9) describing the evolution of $2n - 1$ functions p_i, a_i possesses n Riemann invariants (5.10). System (5.10), for $\beta = 0$, belongs to the class of systems of the hydrodynamic type (Whitham [1965], Dubrovin and Novikov [1989]).

System (5.9) for solutions independent of the variable y is transformed into the famous Toda lattice.

System (5.9) as $\beta = 1/y$ admits the reduction

$$a_i = y^2 \alpha_i(t), \quad p_i = y\rho_i(t) \tag{5.11}$$

and is reduced to the dynamical system

$$\alpha_{k_t} = 2\alpha_k((k+1)\rho_{k+1} - (k-1)\rho_{k-1}), \tag{5.12}$$

$$\frac{1}{2}\rho_{k_t} = \rho_k^2 + (2k+1)\alpha_k - (2k-3)\alpha_{k-1},$$

derived in Bogoyavlenskij [1990]. System (5.12) is equivalent to the operator equation

$$L_t = 2L^2 + [L, A] \tag{5.13}$$

with the same matrices L and A (5.1) and (5.2).

II. System (5.9) in the new coordinates $u_1, \ldots, u_{2n-1}$, defined by the formulae $p_i = u_i$, $a_j = u_{n+j}$, $j = 1, \ldots, n-1$, assumes the form

$$u_{i_t} = \sum_{j=1}^{2n-1} A^{ij} \partial H/\partial u_j, \tag{5.14}$$

$$H = \frac{1}{2}\operatorname{Tr} L^2 = \frac{1}{2}(p_1^2 + \cdots + p_n^2) + a_1 + \cdots + a_{n-1}. \tag{5.15}$$

The operators A^{ij} are determined by the formulae

$$A^{ij} = g^{ij}\frac{\partial}{\partial y} + \sum_{k=1}^{2n-1} b_k^{ij} u_{k_y} + \beta I^{ij}. \tag{5.16}$$

Here the coefficients g^{ij} are symmetric, the coefficients I^{ij} are skew-symmetric, and are non-zero only in the following cases $(1 \le i \le n)$

$$g^{ii} = \frac{2}{3}p_i, \quad g^{n+i,\,i} = g^{i,\,n+i} = g^{n+i,\,i+1} = g^{i+1,\,n+i} = a_i,$$

$$I^{i,\,n+i} = -I^{n+i,\,i} = I^{n+i,\,i+1} = -I^{i+1,\,n+i} = a_i,$$

$$b_i^{ii} = \frac{1}{3}p_i, \quad b_{n+k}^{i,\,n+i} = b_{n+k}^{n+i,\,i+1} = \frac{a_i}{a_k} \quad (1 \le k \le i),$$

$$b_{n+k}^{i+1,\,n+i} = b_{n+k}^{n+i,\,i} = -\frac{a_i}{a_k} \quad (1 \le k \le i-1). \tag{5.17}$$

It is simple to check that the following equalities are valid

$$\partial g^{ij}/\partial u_k = b_k^{ij} + b_k^{ji}. \tag{5.18}$$

Therefore the Christoffel symbols $\Gamma^i_{jk} = -g_{jm}b_k^{mi}$ determine the non-symmetric differential-geometric connection which is compatible with the metric g^{ij}. In view of the relations (5.18) the scalar product (2.34) with the operator coefficients A_{ij} given by (5.16) is skew-symmetric.

III. Let us indicate another form of the system (5.9). After the substitution

$$a_i = \exp(q_{i+1} - q_i) \tag{5.19}$$

we get from the equation (5.7)

$$-b_i = (q_i + q_{i+1})_y + 2\beta, \qquad x_i = b_i \exp(q_{i+1} - q_i). \tag{5.20}$$

Equations (5.5) and (5.6) after substituting (5.19) and (5.20) assume the form

$$p_{i_t} = 2p_i p_{i_y} + 4(q_{i+1,y} + \beta)e^{2(q_{i+1} - q_i)} - 4(q_{i-1_y} + \beta)e^{2(q_i - q_{i-1})}, \tag{5.21}$$

$$q_{i_t} = 2p_i q_{i_y} + p_{i_y} + 2 \sum_{k=1}^{k=i-1} p_{k_y} + 2\beta p_i + \gamma,$$

where β and γ are arbitrary functions of t, y. Each solution of the system (5.21), after the substitutions (5.19) and (5.20), determines a solution of equations (5.5) and (5.6); for concreteness we suppose $\gamma = 0$.

System (5.21) for solutions independent on the variable y coincides with the standard form of the Toda lattice. Equations (5.21) as $\beta = \gamma = 0$ define an elegant system of hydrodynamic type, possessing n Riemann invariants (5.10).

IV. The operator equation (5.3) coincides with the Lax equation

$$L_t = [L, A - L\partial_y - \partial_y L]. \tag{5.22}$$

Therefore the constructed systems of equations (5.9) and (5.21) are equivalent also to the Lax equation (5.22). Systems (5.9) and (5.21) are connected with the system (2.9) in the same way as Toda lattice is connected with Volterra model. Indeed, it is simple to check that from the operator equation (2.3) for the matrix L the equation for matrix $L_1 = L^2$ follows

$$2L_{1_t} = L_1 L_{1_y} + L_{1_y} L_1 + [L_1, A_1], \tag{5.23}$$

where $A_1 = 2A - [L, L_y]$. Equation (5.23) obviously coincides with the equation (5.3). The matrix L_1 is symmetric and has only the following non-zero entries $L_{1_{ii}}$, $L_{1_{i,i+2}} = L_{1_{i+2,i}}$. The matrix A_1 is skew-symmetric and has only the following non-zero entries $A_{1_{i,i+2}} = -A_{1_{i+2,i}}$. Therefore the system (2.9) subjected to the transformation $L_1 = L^2$ is converted onto an invariant submanifold of the system (5.9) and (5.21).

6 Continuous limit

I. We assume that there exist the smooth functions $p(t, x, y)$, $a(t, x, y)$ such that

$$p_j(t, y) = p(t, x_j, y), \qquad a_j(t, y) = a(t, x_j, y), \qquad x_j = j\varepsilon \tag{6.1}$$

and $\beta = \beta_0 \varepsilon^{-1}$. After the substitution of (6.1) and in the limit $\varepsilon \to 0$, equations (5.9) are transformed into the following equations:

$$p_t = pp_y + 2a_y + a_x \int_0^x a_y a^{-1}(t, \xi, y)d\xi + \beta_0 a_x, \tag{6.2}$$

$$a_t = pa_y + 2ap_y + \beta_0 a p_x + a p_x \int_0^x a_y a^{-1}(t, \xi, y)d\xi.$$

Hence, after the substitution $a = \exp u_x$ we get

$$p_t = pp_y + (e^{u_x} u_y)_x + (e^{u_x})_y + \beta_0 (e^{u_x})_x, \tag{6.3}$$

$$u_t = pu_y + 2\int_0^x p_y(t,\xi,y)d\xi + \beta_0 p.$$

Equations (6.2) to (6.3) are the continuous limit of the hydrodynamic type system (5.9).

Equations (6.3) for solutions independent on the variable y are transformed into the equation

$$u_{tt} = \beta_0(e^{u_x})_x. \tag{6.4}$$

This dispersionless limit of the Toda lattice was studied by Brockett and Bloch [1990], where a countable family of its first integrals was derived and an application of (6.4) as a sorter was found. Hamiltonian structures of this equation were investigated by Bloch, Brockett, Kodama and Ratiu [1990] and by Gümral and Nutku [1990]. Hodograph transformation was applied by Kodama [1988], and by Kodama and Gibbons [1989], for studying various dispersionless equations.

Equation (6.4) and the dispersionless limit of the two-dimensional Toda lattice

$$u_{ty} = (e^{u_x})_x \tag{6.5}$$

were derived independently by Bogoyavlenskij [1987]. Equation (6.4) is transformed by using the hodograph transformation into the linear equation

$$v_{rr} = e^r v_{pp}. \tag{6.6}$$

The solution of (6.6) by the Fourier method obviously is reduced to the solution of the Riccati equation $\psi_r + \psi^2 = \exp(r)$. A countable family of first integrals for (6.4) was also obtained in Bogoyavlenskij [1988]; the Proposition 6.1 (see below) generalizes this result for $2+1$-dimensional system (6.2).

The hodograph transformation was applied for the investigation of dispersionless limits of nonlinear systems for the first time by Zabusky [1962], in work devoted to the Fermi-Pasta-Ulam problem.

II. Let us suppose that the functions $|p(t,x,y)|$ and $|a(t,x,y)|$ tend to zero rapidly enough as $|x|,\ |y| \to \infty$.

Proposition 6.1 *The system of equations (6.2) has a countable family of first integrals $(m = 1, 2, \dots)$*

$$I_m = \sum_{k=0}^{[m/2]} \frac{m!}{(m-2k)!k!} \int_{-\infty}^{\infty} p^{m-2k}(t,x,y)a^k(t,x,y)dxdy. \tag{6.7}$$

Proof In view of Lemma 1.7, the system (5.9) has a countable family of conservation laws (1.28) and (1.29). In the continuous limit (6.1), the function $Tr(L^m)$ for the matrix L given by (5.1) is transformed into the integral with respect to x of the coefficient c_m of E^0 in the expansion of the function

$$(E\tilde{a}^{1/2}(t,x,y) + p(t,x,y) + E^{-1}\tilde{a}^{1/2}(t,x,y))^m \tag{6.8}$$

into a power series in the parameter E. Here the substitution $\tilde{a}_i = a_i^2$ is used (see the derivation of the system (5.9)). The coefficient c_m is defined as

$$c_m = \sum_{k=0}^{[m/2]} \frac{m!}{(m-2k)!(k!)^2}p^{m-2k}\tilde{a}^k. \tag{6.9}$$

Therefore the limit value of the integral (1.29) is defined by the formula (6.7). $\square$

7 Integrable extension of the hydrodynamic type system, connected with the Toda lattice

Let us consider a Lax equation of the form

$$(L + \alpha\partial_y)_t = [L + \alpha\partial_y, \gamma\alpha^{-1}L^2 + A], \tag{7.1}$$

where the matrices L and A have only the following non-zero entries

$$L_{i,i+1} = L_{i+1,i} = a_i, \qquad L_{ii} = p_i, \tag{7.2}$$

$$A_{i,i+1} = -A_{i+1,i} = x_i, \qquad A_{ii} = z_i. \tag{7.3}$$

Equation (7.1) is equivalent to the matrix equation

$$L_t = \gamma(LL_y + L_yL) + \alpha A_y + [L, A] \tag{7.4}$$

and after the substitution of the formulae (7.2) and (7.3), is reduced to the system of equations

$$\gamma(a_i a_{i+1})_y = x_i a_{i+1} - a_i x_{i+1}, \tag{7.5}$$

$$\alpha x_{iy} = (z_i - z_{i+1})a_i, \tag{7.6}$$

$$p_{it} = \gamma(p_i^2 + a_i^2 + a_{i-1}^2)_y + 2a_{i-1}x_{i-1} - 2a_i x_i + \alpha z_{iy}, \tag{7.7}$$

$$a_{it} = \gamma(a_i(p_i + p_{i+1}))_y + x_i(p_i - p_{i+1}). \tag{7.8}$$

Solutions of equations (7.5), after the substitution

$$x_i = a_i b_i, \qquad a_i = \exp(q_{i+1} - q_i), \tag{7.9}$$

have the form

$$b_i = -\gamma(q_i + q_{i+1})_y - 2\beta, \quad x_i = b_i\exp(q_{i+1} - q_i). \tag{7.10}$$

Solutions of equations (7.6) are determined by the formulae

$$z_k = 2\alpha\beta q_{ky} + \alpha\gamma(q_{ky})^2 + 2\alpha\gamma\sum_{j=1}^{k-1} q_{jyy} + \alpha\gamma q_{kyy}. \tag{7.11}$$

Equations (7.7) and (7.8) after substituting the obtained formulae (7.10) and (7.11), assume the form

$$p_{kt} = 2\gamma p_k p_{ky} + 4(\gamma q_{k+1,y} + \beta)\exp^{2(q_{k+1} - q_k)} - 4(\gamma q_{k-1,y} + \beta)\exp^{2(q_k - q_{k-1})}$$

$$+ 2\alpha^2\beta q_{kyy} + 2\alpha^2\gamma q_{ky}q_{kyy} + 2\alpha^2\gamma\sum_{j=1}^{k-1} q_{jyyy} + \alpha^2\gamma q_{kyyy},$$

$$q_{kt} = 2\gamma p_k q_{ky} + \gamma p_{ky} + 2\gamma\sum_{j=1}^{k-1} p_{jy} + 2\beta p_k. \tag{7.12}$$

System (7.12) in view of the above considerations admits the Lax representations (7.1) and is transformed into hydrodynamic type system (5.21) as $\alpha = \beta = 0$. System (7.12) for $\gamma = 0$ coincides with the two-dimensional Toda lattice (Mikhailov [1979])

$$p_{kt} = 4\beta e^{2(q_{k+1} - q_k)} - 4\beta e^{2(q_k - q_{k-1})} + 2\alpha^2\beta q_{kyy},$$

$$q_{kt} = 2\beta p_k. \tag{7.13}$$

System (7.12) admits the equivalent Lax representation

$$(L + \alpha\partial_y)_t = [L + \alpha\partial_y, A - \gamma(\partial_y L + L\partial_y) - \gamma\alpha\partial_y^2], \qquad (7.14)$$

which remains regular in the continuous limit $\alpha \to 0$, unlike the Lax representation (7.1).

References

Bloch, A.M., Brockett, R.W., Kodama, Y., and Ratiu, T. [1990] *Spectral equation for the long wave limit of the Toda lattice equations*, Proceedings of the CRM Workshop on Hamiltonian systems, transformation groups and spectral transform methods. (J. Harnad and J.E. Marsden, eds.), CRM, Montreal.

Bogoyavlenskij, O. I. [1991] *Breaking solitons. Nonlinear integrable equations*, Nauka, Moscow.

______ [1989] *Breaking solitons in new two-dimensional integrable equations*, zv. Akad. Nauk SSSR ser. Mat., **53**, 243–257; English transl. in Math. USSR Izv. **34** (1990), 243–257.

______ [1990] *Breaking solitons in 2+1-dimensional integrable equations*, Uspekhi Mat. Nauk., **45:4**, 17–77; English transl. in Russian Math. Surveys **45:4** (1990), 1–86.

______ [1987] *Some constructions of integrable dynamical systems*, Izv. Akad. Nauk SSSR Ser. Mat., **51**, 737–766; English transl. in Math. USSR Izv. **31** (1988), 47–75.

______ [1988] *Lax representation with spectral parameter for some dynamical systems*, Izv. Akad. Nauk SSSR Ser. Mat., **52**, 243–266; English transl. in Math. USSR Izv. **32** (1989), 245–268.

Brockett, R.W., and Bloch, A.M. [1990] *Sorting with the dispersionless limit of the Toda lattice*, Proceedings of the CRM Workshop on Hamiltonian systems transformation groups and spectral transform methods, (J. Harnad and J.E. Marsden, eds.), CRM, Montreal.

Calogero, F., and Degasperis, A. [1976] *Non-linear evolution equations solvable by the inverse spectral transform, I* Nuovo Cimento B, **(11) 32**, 201–242.

______ [1977] *Non-linear evolution equations solvable by the inverse spectral transform, II* Nuovo Cimento B, **(11) 39**, 1–54.

______ [1980] *Non-linear evolution equations solvable by the inverse spectral transform associated with the matrix Schrodinger equation*, Solitons Springer-Verlag, Berlin - New York.

Dodd, R.K., Ellbeck, J.C., Gibbon, J.D., and Morris, H.C. [1982] *Solitons and non-linear wave equations*, Academic Press, London.

Dubrovin, B.A., and Novikov, S.P. [1989] *Hydrodynamics of weakly deformed soliton lattices. Differential geometry and Hamiltonian theory*, Uspekhi Mat. Nauk **44**, 29–98; English transl. in Russian Math. Surveys 44 (1989).

Gümral, H., and Nutku, Y. [1990] *Multi-Hamiltonian structure of equations of hydrodynamic type*, preprint, Bilkent University.

Kodama, Y. [1988] *A method for solving the dispersionless KP equation and its exact solutions*, Physics Letters A, **129**, 223–226.

Kodama, Y., and Gibbons, J. [1989] *A method for solving the dispersionless KP equation and its exact solutions, II*, Physics Letters A, **135**, 167–170.

Mikhailov, A.V. [1979] *On the integrability of the two-dimensional generalization of the Toda lattice*, Pis'ma Zh. Exp. Teor. Fiz., **30**, 443–448; English transl. in Soviet Physics JETP Letters **30** (1979), 414–418.

Moser, J. [1975] *Three integrable Hamiltonian systems connected with isospectral deformations*, Advanc. in Math., **16**, 197–220.

Whitham, G.B. [1965] *Non-linear dispersive waves*, Proc. Royal Soc., **A283**, London, 238–261.

Zabusky, N. [1962] *Exact solutions for the vibrations of a nonlinear continuous model string*, Journal of Mathematical Physics, **3**, 1028–1039.

Zakharov, V.E. [1983] *The method of inverse scattering problem*, Appendix to; Russian translation,R.K. Bullough and P.J. Caudrey (eds.), Solitons, Mir, Moscow.

Fields Institute Communications
Volume **3**, 1994

The Double Bracket Equation as the Solution of a Variational Problem

Roger Brockett
Division of Applied Sciences
Harvard University
Cambridge, MA, USA
02138

Abstract. We consider a class of optimal control problems whose solution provides a variational interpretation of the Lie algebraic differential equation $\dot{x} = [x, [x, n]]$ and its various specialization's such as the Toda lattice equations investigated in Bloch, Brockett and Ratiu [1990]. As a result, we are lead to a variety of solvable nonlinear optimal control problems and to new connections between optimal control theory and the recent literature on integrable hamiltonian systems.

1 Introduction

Let $\mathfrak{G}$ be a real compact connected Lie group with Lie algebra $\mathfrak{g}$. Let x and u be functions of time, taking on values in $\mathfrak{g}$. For suitably well-behaved choices of u, the differential equation

$$\dot{x} = [x, u] \; ; \;\; x(0) = x_0$$

evolves on the adjoint orbit $\mathfrak{O}(x_0)$ generated by $\mathfrak{G}$ acting on $\mathfrak{g}$ via the standard adjoint action

$$\phi(G, x) = G \frac{d}{dt} \exp(tx)|_{t=0} G^{-1} \; .$$

We can interpret $\dot{x} = [x, u]$ as defining a control system on the orbit $\mathfrak{O}(x_0)$. It is, almost by definition, controllable in the sense that for any given positive time t_f and any given pair of points x_1 and x_2 on the orbit, there exists a control that steers the system from x_1 to x_2 in t_f units of time.

1991 *Mathematics Subject Classification.* Primary 93C10;Secondary 34A05.

This work was supported in part by the National Science Foundation under Engineering Research Center Program, NSF D CRD-8803012, the US Army Research Office under grant DAAL03-92-G-0164 (Center for Intelligent Control Systems), DARPA and the Air Force under contract F49620-92-J-0466, and by the Office of Naval Research under contract N00014-92-J-1887.

Supported in part by the Ministry of Colleges and Universities of Ontario and the Natural Sciences and Engineering Research Council of Canada while visiting The Fields Institute.

Notwithstanding the apparent simplicity of this control system, it gives rise to a rather interesting class of optimization problems. Let x be an element of $\mathfrak{g}$ and let $ad_x(\cdot)$ denote the operator $[x, \cdot]$ mapping $\mathfrak{g}$ into itself. Under the present assumptions on $\mathfrak{g}$ the quadratic form $-tr(ad_x ad_x)$ is positive definite; we find it both suggestive and convenient to abbreviate it as $\|x\|^2$. Consider the problem of minimizing the functional

$$\eta = \int_0^{t_f} \|[x, n]\|^2 + \|u\|^2 \, dt \ .$$

The limiting case obtained by letting t_f be infinite will be of particular interest. We will show that this problem is similar to the standard linear quadratic problem that plays a central role in the application of control theory.

The control theoretic context for this type of question is the so-called *regulator problem* in which one is given a control system

$$\dot{x} = a(x, u) \ ; \ \ x(0) = x_0$$

and a performance criterion to be minimized

$$\eta = \int_0^{t_f} L(x, u) \, dt \ .$$

If t_f is finite, then the control that minimizes η will typically have an explicit time dependence that can not be eliminated by expressing u as a function of x. If t_f is infinite, then there are conditions that must be met if such problems are to have solutions. In particular, L must be nonnegative and L and a must both vanish at one or more points. Otherwise, the integral can not converge. One special feature of the infinite time regulator problem is that it is often possible to eliminate any explicit time dependence in the control law, thus giving a time-invariant feedback solution to the optimal control problem. If $L(\ ,\ )$ vanishes at x_0, u_0 and at no other point, then $a(\ ,\ )$ must vanish there also, and the optimal control must steer the system to this point. If $L(\ ,\ )$ and $a(\ ,\ )$ vanish at more than one point, then it is necessary to distinguish between those problems for which the final values of x and u are specified in advance and those for which the final value is subject to choice and must be optimized. We will distinguish between these by the terms *fixed end-point regulator problem* and *free end-point regulator problem*.

2 A regulator problem on compact adjoint orbits

If n is an element of $\mathfrak{g}$, let $c(n)$ denote the set of all elements of $\mathfrak{g}$ that commute with n. From now on we assume that x_0 and n are *regular* elements of $\mathfrak{g}$ in the sense that the dimensions of $c(x_0)$ and $c(n)$ are minimal. Given a particular adjoint orbit on which the system evolves, consider the loss function

$$\eta = \int_0^{\infty} \|[x, n]\|^2 + \|u\|^2 \, dt \ .$$

The integrand vanishes if x and n commute and u is zero. Because we have assumed that n and x_0 are regular, there are only a finite number of points at which the integrand vanishes.

In a real compact semi-simple Lie algebra, ad_n maps the orthogonal complement of $c(n)$ into itself, and maps $c(n)$ into zero, we denote by ad_n^{-1} the inverse of the restriction of ad_n to the orthogonal complement of $c(n)$. If we use the differential equation to express u as $\mathrm{ad}_x^{-1}(\dot{x})$, then the optimal control problem can be recast as a standard variational problem

$$\eta = \int_0^\infty \|[x,n]\|^2 + \|\mathrm{ad}_x^{-1}(\dot{x})\|^2 \, dt \ .$$

To obtain the Euler-Lagrange equations corresponding to this problem we note that the derivative of $\|[x,n]\|^2$ with respect to x is $-2\mathrm{ad}_n^2(x)$ and that the derivative of $\|\mathrm{ad}_x^{-1}(\dot{x})\|^2$ with respect to $\dot{x}$ is $-2\mathrm{ad}_x^{-2}(\dot{x})$. To compute the derivative of $\|\mathrm{ad}_x^{-1}(\dot{x})\|^2$ with respect to x we note that for any constant p in $\mathfrak{g}$ the quantity $[x, \mathrm{ad}_x^{-1}(p)]$ is constant and so by the chain rule

$$\frac{d}{dt}\, \mathrm{ad}_x^{-1}(p) = \mathrm{ad}_x^{-1}([\mathrm{ad}_x^{-1}(p), \dot{x}]).$$

This implies that

$$\frac{d}{dt}\, \mathrm{ad}_x^{-2}(p) = \mathrm{ad}_x^{-2}([\mathrm{ad}_x^{-1}(p), \dot{x}]) + \mathrm{ad}_x^{-1}([\mathrm{ad}_x^{-2}(p), \dot{x}]).$$

Of course the right-hand side is just the partial derivative of $\|\mathrm{ad}_x^{-1}(p)\|^2$ with respect to x evaluated in the direction $\dot{x}$. Putting these results together gives the Euler-Lagrange equation

$$\ddot{x} = -[\mathrm{ad}_x^{-1}(\dot{x}), \dot{x}] + \mathrm{ad}_x^2 \mathrm{ad}_n^2(x) \ .$$

This can be compared with (and corrects a sign error appearing in) the geodesic equation on $\mathcal{O}(x_0)$ given in Brockett [1993], page 72. In addition to making use of the Euler-Lagrange equation, we point out a direct argument that facilitates the investigation of the existence of optimal controls. It is based on a completion of the square inside the integral. Specifically, if h is a constant element of $\mathfrak{g}$ and if $\dot{x} = [x, u]$, then

$$\mathrm{tr}(\mathrm{ad}_h \mathrm{ad}_x)\big|_0^\infty = \int_0^\infty \mathrm{tr}(\mathrm{ad}_h \mathrm{ad}_{[x,u]}) \, dt \ .$$

Using this equality we rewrite the given functional as

$$\eta = \mathrm{tr}(\mathrm{ad}_h \mathrm{ad}_x)\big|_0^\infty + \int_0^\infty \|[x,n]\|^2 + \|u\|^2 - 2\mathrm{tr}(\mathrm{ad}_h \mathrm{ad}_x \mathrm{ad}_u) \, dt \ .$$

The critical observation is that if h satisfies the quadratic equation

$$(\mathrm{ad}_h)^2 = (\mathrm{ad}_n)^2 \ ,$$

then we can express η as

$$\eta = \mathrm{tr}(\mathrm{ad}_h \mathrm{ad}_{x(t)})\big|_0^\infty + \int_0^\infty \|[x,h] + u\|^2 \, dt \ .$$

This makes it clear that if the control law $u = [x, -h]$ steers the system to the desired final value, then it is optimal, and that the corresponding cost $\eta^* = \mathrm{tr}(\mathrm{ad}_h \mathrm{ad}_{x_f - x(0)})$ is minimal. Of course, in this case the optimal closed loop system is governed by the equation

$$\dot{x} = [x, [x, -h]] \ .$$

Theorem 2.1 *Let $\mathfrak{g}$ be a real, compact semisimple Lie algebra. Let $\mathfrak{O}(x_0)$ be an adjoint orbit, and let x_f and n be commuting elements of $\mathfrak{O}(x_0)$. If $(\mathrm{ad}_h)^2 = (\mathrm{ad}_n)^2$, then the minimal cost associated with the fixed end-point regulator problem*

$$\dot{x} = [x, u] \; ; \; x(0) = x_0 \; ; \; x(\infty) = x_f$$

$$\eta = \int_0^\infty \|[x, n]\|^2 + \|u\|^2 \, dt$$

is

$$\eta^* = \mathrm{tr}(\mathrm{ad}_h \mathrm{ad}_{x_f - x(0)})$$

and the feedback control law $u = [x, -h]$ achieves this cost, provided that the solution $\dot{x} = [x, [x, -h]]$; $x(0) = x_0$ approaches x_f as t goes to infinity.

Proof We add and subtract $\mathrm{tr}(\mathrm{ad}_h \mathrm{ad}_{x(t)})$ as above. In view of the fact that $[h, x] = 0$ if and only if $\mathrm{tr}([n, x][n, x]) = 0$ we see that the vanishing of $[n, x]$ together with $(\mathrm{ad}_h)^2 = (\mathrm{ad}_n)^2$ implies that $[h, x(\infty)] = 0$. Thus, the lower bound is established. That this is achieved using the control $u = [x, -h]$ is likewise obvious. $\square$

Remark 2.2 All trajectories of the optimal system stay in the subalgebra of $\mathfrak{g}$ generated by $x(0)$ and n. This algebra may or may not be simple. If it is, then the only solutions of $(\mathrm{ad}_h)^2 = (\mathrm{ad}_n)^2$ are the solutions $h = n$ and $h = -n$. In this case, given x_f such that $[n, x_f] = 0$, it is only possible to find h that meets the conditions of Theorem 2.1 if the function $m(x) = \mathrm{tr}(\mathrm{ad}_n \mathrm{ad}_x)$ considered as a function on $\mathfrak{O}(x_0)$ takes on a maximum or minimum at x_f. More generally, there exist solutions of $(\mathrm{ad}_h)^2 = (\mathrm{ad}_n)^2$ that take the form $h = +n$ on one subalgebra and $h = -n$ on a different subalgebra, see Remark 2.3 below.

Remark 2.3 Of course x is constrained to lie on a submanifold of the Lie algebra $\mathfrak{g}$. Consequently, there is a degree of arbitrariness in the expression for the control law and the minimum return function η^*. For example, if x takes on values in the space of skew-hermitian matrices, and if n is a diagonal skew-hermitian matrix, then one could parametrize the adjoint orbit in a neighborhood of an equilibrium point by the off-diagonal elements of x. The control law $[x, -n]$ would, then, not depend on the diagonal elements but the given expression for the minimum cost function $\eta*$ is entirely in terms of the diagonals. Re-expressing it in terms of the off diagonals, while possible, would not yeild a simple expression.

3 Stable manifolds and linearization

Usually it will be impossible to find a solution of the given quadratic equation for h that makes x_f an asymptotically critical point. In those cases where this is impossible, one expects that the optimal control will still exist, at least for most initial conditions. To explore this we investigate a linearization technique valid near any equilibrium point.

Let x_f and n be commuting elements of $\mathfrak{O}(x(0))$. If we linearize the system $\dot{x} = [x, u]$ about the solution $x(t) = x_f$ and let $w = x - x_f$, then w satisfies the equation

$$\dot{w} = [x_f, u] \; ; \; w(0) = x_0 - x_f \, .$$

This equation is not controllable on all of $\mathfrak{g}$ because x itself is constrained to lie on the adjoint orbit $\mathfrak{O}(x(0))$. The reachable set from $w(0) = 0$ is the orthogonal complement of $c(x_f)$. We seek to minimize the integral

$$\eta = \int_0^\infty \|[w + x_f, n]\|^2 + \|u\|^2 \, dt$$

or, equivalently,

$$\eta = \int_0^\infty \|[w, n]\|^2 + \|u\|^2 \, dt .$$

A standard application of optimal control theory gives a steady state Riccati equation of the form

$$\mathrm{ad}_k \mathrm{ad}_{x_f}^2 \mathrm{ad}_k = \mathrm{ad}_n^2 .$$

The product $\mathrm{ad}_n \mathrm{ad}_{x_f}$ is necessarily self adjoint because it is the product of two commuting skew-adjoint operators. Its eigenvalues are the products of the squares of purely imaginary quantities and so are positive. This equation can be solved by pre and post multiplying by $\mathrm{ad}_{x_f}^{-1}$, taking a positive square root, and then pre and post multiplying by $\mathrm{ad}_{x_f}^{-1}$. The result is

$$\mathrm{ad}_k = \mathrm{ad}_{x_f}^{-1} (\mathrm{ad}_n^2 \mathrm{ad}_{x_f}^2)^{1/2} \mathrm{ad}_{x_f}^{-1} .$$

The optimal trajectories for the linearized system satisfy

$$\dot{w} = [x_f, [x_f, [k, w]]]$$

or, equivalently,

$$\dot{w} = -(\mathrm{ad}_n^2 \mathrm{ad}_{x_f}^2)^{1/2}(w) .$$

It is remarkable that the optimal control for the linearized system, when used to control the original nonlinear system, remains optimal.

Remark 3.1 A direct linearization of the Euler-Lagrange equation leads to

$$\ddot{w} = \mathrm{ad}_{x_f}^2 \mathrm{ad}_n^2 (w) .$$

There is a stable submanifold associated with the equilibrium point that is of the same dimension as that of x.

The minimal cost associated with the linearized problem can be worked out from the steady state solution of a Riccati equation. This yields

$$\eta^* = \mathrm{tr}(\mathrm{ad}_w \mathrm{ad}_{[k,w]}) .$$

Remark 3.2 Let π^+ denote the projection operator that projects $\mathfrak{g}$ (orthogonally with respect to the Killing form) onto the invariant subspace of $\mathrm{ad}_n \mathrm{ad}_{x_f}$ on which it has positive eigenvalues and let π^- denote the projection operator that projects $\mathfrak{g}$ onto the invariant subspace of $\mathrm{ad}_n \mathrm{ad}_{x_f}$ on which it acts with negative eigenvalues. It is clear that on the stable submanifold of the equation $\dot{x} = [x, [x, n]]$ that passes through x_f the optimal control is $u = [x, n]$. On the other hand, on the stable manifold of $\dot{x} = -[x, [x, n]]$ passing through x_f the control $u = -[x, n]$ is optimal. Because the elements n and x_f are regular, the equilibrium point has no zero eigenvalues in the linearization, and so the sum of these two spaces span the tangent space of $\mathfrak{O}(x_0)$ at x_f.

4 Symmetric spaces

If $\mathfrak{g}$ admits an automorphism ι that is of period two, i.e. if ι^2 is the identity, then $\mathfrak{g}$ can be expressed as the vector space sum of the plus one eigenspace of ι and the minus one eigenspace of ϕ. In this case we can restrict u to the positive eigenspace and restrict x to the negative eigenspace, and the above variational theory carries through unchanged. The special unitary group, unitary matrices having determinant one, gives rise to an example. In this case the Lie algebra is the set of n by n skew-hermitian matrices having zero trace. The automorphism is the map that sends a matrix into its conjugate. The plus one eigenspace is the (real) skew-symmetric matrices and the minus one eigenspace is the set of purely imaginary (necessarily symmetric) matrices. The basic control problem with x in the minus one eigenspace and u in the plus one eigenspace then gives rise to a situation in which the above analysis applies.

Example 4.1 Let $\mathfrak{O}(\lambda_1, \lambda_2, \dots, \lambda_n)$ denote the set of all real n by n symmetric matrices with eigenvalues $\{\lambda_1, \lambda_2, \dots, \lambda_n\}$. Consider

$$\dot{H} = [H, U] \ : \ H(0) \in \mathfrak{O}(\lambda_1, \lambda_2, \dots, \lambda_n) \,,$$

with $U = -U^T$. Let N be a diagonal element of $\mathfrak{O}(\lambda_1, \lambda_2, \dots, \lambda_n)$, say $N = \mathrm{diag}(n_{11}, n_{22}, \dots, n_{nn})$. Consider the problem of minimizing

$$\int_0^\infty -\mathrm{tr}[H, N]^2 - \mathrm{tr}\, U^2 \, dt$$

subject to $H(\infty) = D = \mathrm{diag}(\lambda_{\pi(1)}, \lambda_{\pi(2)}, \dots \lambda_{\pi(n)})$. Using Theorem 2.1 we see that if D and N are similarly ordered or oppositely ordered, then the control law $U = [H, N]$ or $-[H, N]$ is optimal. Otherwise, we need to resort to linearization. If we linearize near D and introduce $W = H - D$, then

$$\dot{w}_{ij} \ = \ (d_i - d_j)\, u_{ij}$$
$$\eta \ = \ \int_0^\infty \sum\sum (n_i - n_j)^2 w_{ij}^2 + u_{ij}^2 \, dt \,.$$

The optimal control for the linearized system is

$$u_{ij} = -\sqrt{\frac{(n_i - n_j)^2}{(d_i - d_j)^2}}\,(d_i - d_j)\, w_{ij}$$

and the minimum cost is

$$\eta = \sum\sum w_{ij}^2 \frac{|n_i - n_j|}{|d_i - d_j|} \,.$$

This is to be compared with the special case when there is similar ordering and

$$\eta = \sum (h_{ii} - d_{ii}) n_{ii} \,.$$

Of course there is a relationship between $\{h_{ii}\}$ and $\{h_{ij}\}$ on an adjoint orbit. In fact,

$$H = e^\Omega D e^{-\Omega} = D + [\Omega, D] + \frac{1}{2}[\Omega, [\Omega, D]] + \cdots \,,$$

for some skew-symmetric matrix $\Omega = (\omega_{ij})$. Thus,

$$h_{ii} = d_{ii} + \frac{1}{2}\sum_k \omega_{ik}^2(d_i - d_k)$$

$$h_{ij} = \omega_{ij}(d_i - d_j) + \frac{1}{2}\sum_k \omega_{ik}\omega_{kj}(d_i + d_j - 2d_k)\,.$$

Squaring the last relationship we see that

$$h_{ij}^2 = w_{ij}^2(d_i - d_j)^2 + \text{cubic terms},$$

thus, we have the approximation

$$h_{ii} = d_{ii} + \sum \frac{1}{2}\left(\frac{h_{ij}}{d_i - d_j}\right)^2(d_i - d_j)\,.$$

Using this in the formula for the cost $\eta = \sum h_{ii}n_{ii}$ we get the approximate result

$$\eta = \sum\sum \frac{h_{ij}^2|n_i - n_j|}{|d_i - d_j|}\,.$$

The feedback control law is

$$u_{ij} = -\sqrt{\frac{(n_i - n_j)^2}{(d_i - d_j)^2}}\,(d_i - d_j)\,w_{ij}\,.$$

If $(n_i - n_j)/(d_i - d_j)$ is positive for all i larger than j this can be simplified to

$$u_{ij} = (n_i - n_j)\,w_{ij}$$

and hence

$$\dot{w}_{ij} = -(n_i - n_j)(d_i - d_j)\,w_{ij}\,.$$

5 The Toda regulator

One of the main points of Bloch, Brockett and Ratiu [1990] is to put in a more general context the observation of Brockett [1993] that if one chooses n properly, then Flaschka's form of the Toda lattice equations can be expressed as a double bracket equation restricted to a certain invariant submanifold. In the present context this can be interpreted as defining a particular class of nonlinear control problems for which the regulator problem can be solved explicitly. We illustrate with the most familiar case, the "standard" (sl(n)), Toda lattice.

Example 5.1 Let $\Omega(n)$ denote the set of n by n skew-symmetric matrices and let $\mathfrak{O}_T(\lambda_1, \lambda_2, \dots, \lambda_n)$ denote the the subset of $\mathfrak{O}(\lambda_1, \lambda_2, \dots, \lambda_n)$ consisting of those elements that are tridiagonal. Assume from now on that the eigenvalues are unrepeated. We can define a controllable system on $\mathfrak{O}_T(\lambda_1, \lambda_2, \dots, \lambda_n)$ in the following way. First of all, define a rank n-1 vector bundle E over $\mathfrak{O}(\lambda_1, \lambda_2, \dots, \lambda_n)$ consisting of the elements of the trivial bundle $\mathfrak{O}_T(\lambda_1, \lambda_2, \dots, \lambda_n) \times \Omega(n)$ such that $[H, \Omega]$ is tridiagonal. Notice that if H_0 is diagonal the set of possible values for Ω is just the $n - 1$ dimensional space of n by n, tridiagonal, skew-symmetric matrices. If $H = \Theta^T D\Theta$ with D diagonal then the fiber over H is just the $n - 1$ dimensional vector space $\Theta^T U\Theta$ with U being tridiagonal. Of course if we restrict H to be tridiagonal we get a rank n-1 vector bundle over $\mathfrak{O}_T(\lambda_1, \lambda_2, \dots, \lambda_n)$. Denote this vector bundle by E'. The equation of motion for the control system of interest here is $\dot{H} = [H, U]$ as in Example 4.1, but now we restrict U, to be such that the pair (H, U) belongs to E'. Matters being so, H will remain tridiagonal if it is initially tridiagonal and thus we have a control system on the space $\mathfrak{O}_T(\lambda_1, \lambda_2, \dots, \lambda_n)$. It is not difficult to see that this system is controllable on the space $\mathfrak{O}_T(\lambda_1, \lambda_2, \dots, \lambda_n)$.

The performance measure to be minimized is of the same form as that considered in Example 4.1 but with the special choice

$$N = diag(1, 2, 3, ..., n).$$

From Example 4.1, we see that if there were no constraints on U then the optimal control would be $U = [H, N]$. However, this particular choice lies in the fiber over H and thus it is also the optimal control for the problem defined on $\mathfrak{O}_T(\lambda_1, \lambda_2, \dots, \lambda_n)$. If the desired end point is characterized by $\lambda_1 \geq \lambda_2 \geq \dots \geq \lambda_n$, then the optimal control law is obtained by specializing the formula $u = [x, n]$. Letting $x_1, x_2, ..., x_n$ denote the diagonal elements of H and $y_1, y_2, ..., y_{n-1}$ denote the elements on the superdiagonal we can express the closed loop equations in the standard Toda lattice form

$$
\begin{aligned}
\dot{x}_1 &= 2y_1^2 \\
\dot{x}_2 &= -2y_1^2 + 2y_2^2 \\
\dot{x}_3 &= -2y_2^2 + 2y_3^2 \\
&\cdots \\
\dot{x}_n &= -2y_{n-1}^2 \\
\dot{y}_1 &= -2(x_1 - x_2)y_1 \\
\dot{y}_2 &= -2(x_2 - x_3)y_2 \\
\dot{y}_3 &= -2(x_3 - x_4)y_3 \\
&\cdots \\
\dot{y}_{n-1} &= -2(x_{n-1} - x_n)y_{n-1} \, .
\end{aligned}
$$

Thus, we see that insofar as this variational problem is concerned, $\mathfrak{O}_T(\lambda_1, \lambda_2, \dots, \lambda_n)$ relates to $\mathfrak{O}(\lambda_1, \lambda_2, \dots, \lambda_n)$ in somewhat the same way that a totally geodesic submanifold relates to the manifold in which it is imbedded.

6 Acknowledgements

The author would like to thank Anthony Bloch and Peter Crouch for their comments on an earlier draft of the manuscript.

References

Bloch, A.M., Brockett, R.W., and Ratiu, T. [1990] *A New Formulation of the Generalized Toda Lattice Equations and their Fixed Point Analysis*, Bul. of the American Math. Soc., **23** no. 2, 447–485.

Brockett, R.W. [1993] *Differential Geometry and the Design of Gradient Algorithms*; in Differential Geometry, **1**, (R. Green and S.-T. Yau, eds.), American Math. Soc., Providence, 69–91.

Bloch, A.M. [1990] *Steepest Descent, Linear Programming and Hamiltonian Systems*, Contemporary Mathematics, American Math. Soc., **114**, 77–88.

Fields Institute Communications
Volume **3**, 1994

Integration and Visualization of Matrix Orbits on the Connection Machine

Jean-Philippe Brunet
Thinking Machines Corporation
245 First Street
Cambridge, MA, USA
02142

Abstract. We numerically integrate Hamiltonian equations of motion for isospectral flows on the space of real symmetric matrices. Integrals of motion that characterize the co-adjoint orbits are preserved. The isospectral deformations of a Toeplitz matrix on its orbit are visualized on the Connection Machine and the concept of abstract matrix factorization is illustrated.

1 Introduction

The connection between matrix factorization and integrable dynamical systems allows one to look at algebraic processes from a different perspective. This is particularly true for the algebraic eigenvalue problem. Since the eigenvalues of a matrix A are the roots of the characteristic polynomial $Det(A - \lambda I) = 0$, they can not be determined directly for $n > 4$, after Galois' theorem. Indeed most eigenvalue algorithms, such as QR or the Jacobi method, iteratively reduce the matrix to diagonal form using a sequence of similarity transformations. Such algorithms can be thought of as sampling a smooth manifold of matrices similar to A. Indeed, there exist matrix dynamical systems that interpolate in time the iterative dynamics of matrix factorization algorithms (Chu [1992], Watkins [1984], Deift, Li and Tomei [1989]). Most notably, the Toda flow (a dynamical system of particles interacting through an exponential potential) is a continuous realization of the QR algorithm to compute eigenvalues of a symmetric tridiagonal matrix (Symes [1982]). The equations of motion are Hamiltonian and the resulting dynamical system is integrable. The 'motion' of the matrix takes place on a smooth manifold: the co-adjoint orbit of the lower triangular group on the dual of its Lie algebra. The tridiagonal form is preserved. The stable equilibrium point of the flow is a diagonal matrix with the eigenvalues exposed on the diagonal in a non-increasing order (for positive time).

1991 *Mathematics Subject Classification.* Primary 70H05;Secondary 15A18.

Supported in part by the Ministry of Colleges and Universities of Ontario and the Natural Sciences and Engineering Research Council of Canada while visiting The Fields Institute.

Diagonal matrices with different orderings of the eigenvalues on the diagonal are saddle points (Deift, Nanda and Tomei [1983]) [1]. The integrals of motion are the eigenvalues of the matrix chosen as initial condition. They account for the complete integrability of the flow.

The mathematical framework is in fact very general and provides a dynamical formulation of the algebraic eigenvalue problem. Symes [1980] provides a detailed account of the theory, original references, and examples. Deift et al., [1986] characterizes in great detail matrix orbits for both symmetric (Deift et al., [1986]) and nonsymmetric (Deift, Li and Tomei [1989]) matrices. The expressions of the invariants that characterize matrix orbits are in term of ratios of polynomials of matrix entries.

Our goal is to integrate matrix orbits and visualize isospectral deformations for some interesting class of matrices. We shall mostly be concerned with the symmetric case throughout the paper. In Section 2, we construct maps that approximate the flow to arbitrary order and preserve the orbit invariants. In Section 3, we suggest that massively parallel processors with distributed memory, such as the Connection machine systems, are well suited for numerically integrating and visualizing matrix orbits in real time. As an illustration, we examine the orbit of a Toeplitz matrix by showing snapshots taken at various stages of its dynamical evolution. We also illustrate the concept of abstract matrix factorization (Chu and Norris [1988]) in Section 4.

It should be stressed that explicit integration of matrix orbits is not a viable alternative to solving the symmetric eigenvalue problem, even on parallel architectures. Instead, Householder reduction to tridiagonal form followed by parallel bisection, or the Jacobi method, can be implemented very efficiently on distributed memory architectures such as the Connection Machine systems (CMSSL release notes [1993]).

2 Integration of matrix orbits

Hamilton equations for the eigenvalue problem can be written as

$$\dot{A} = [A, \Pi_l(A)] \tag{2.1}$$

$$\Pi_l(A) = A_+ + A_0 + A_-^T$$

where A_-, A_0 and A_+ are the strictly lower, diagonal, and strictly upper parts of matrix A.

An alternate set of equations are:

$$\dot{A} = [\Pi_k(A), A] \tag{2.2}$$

with

$$\Pi_k(A) = A_- - A_-^T.$$

Those equations can be interpreted as the infinitesimal (co-adjoint) action of a matrix Lie group on the dual of its Lie algebra identified with the space of symmetric matrices.

---

[1] It is possible to enforce convergence to an arbitrary ordering of the eigenvalues using gradient flows (Brockett [1991], Chu [1992]), but the tridiagonal form is not preserved anymore.

Hamilton equation (2.1) is the infinitesimal version of the co-adjoint action of a subgroup of $L^+(n, R)$, the lower triangular group with positive diagonal entries, on the dual of its Lie algebra, l, while equation (2.2) involves the orthogonal group and its Lie algebra, k, the space of skew-symmetric matrices. $\Pi_l(A)$ and $\Pi_k(A)$ are members of the algebras l and k, respectively. The duality of Hamilton equations (2.1, 2.2) reflects the natural decomposition of the Lie algebra of the linear group into the algebras of lower triangular and skew symmetric matrices. The concomitant decomposition of the linear group is the well known QL factorization.

In order to integrate equations (2.1, 2.2), we construct a map that approximates the flow generated by those equations up to some prescribed order p in time. The construction of high order maps requires the evaluation of higher derivatives of the matrix A, which in turn requires the evaluation of high derivatives of the Lie algebra elements. But this is easily accomplished using commutator rules and the fact that the derivative of the Lie algebra element $\Pi_k(A)$ is the Lie algebra element evaluated at the derivative $\Pi_k(\dot{A})$. For instance, the second derivative $\ddot{A}$ is computed as

$$\ddot{A} = [\dot{\Pi}_k(A), A] + [\Pi_k(A), \dot{A}] = [\Pi_k(\dot{A}), A] + [\Pi_k(A), \dot{A}].$$

Following such rules, it is straightforward to construct the *pth* order map $A_i \rightarrow A_{i+1}$ for time step h

$$A_{i+1} = A_i + \left(\frac{h^k}{k!}\right) \sum_{k=1}^{p} \frac{d^k(A_i)}{dt} + O(h^{p+1}).$$

Given initial conditions, i.e. a matrix A_0, one can approximate to arbitrary order the trajectory of the flow generated by Hamilton equations (2.1, 2.2) by successive applications of the above map.

Direct numerical evaluation of such maps is rather costly. For example the second order map requires three commutator evaluations. In order to reduce the computation cost, we can use a *pth* order explicit Runge-Kutta scheme to evaluate the *pth* order map. In the numerical experiments reported below, we use a *5th* order Runge-Kutta-Felhberg scheme which only requires six commutator evaluations and provides an estimate of the truncation error used to adjust the step size (Press and Teukolsky [1992]).

The above *pth* order map preserves the integrals of motion up to order p. Such integrals of motion are integral invariants of the matrix (co-adjoint) orbit under the Hamiltonian flow. The co-adjoint orbit of L^+ through a (generic) matrix A has been characterized in great detail by Deift et al., [1986]. Besides the n eigenvalues of the matrix which are preserved by the flow there are $n(n-2)/2$ additional conserved quantities (for n even). Those additional co-adjoint invariants are ratios of polynomials of matrix entries. When the orbit approaches the fixed point of the flow, the matrix is very close to diagonal form, both the numerator and denominator of the co-adjoint invariants tend to zero and it is remarkable that the ratios would remain constant. We checked (on small systems and using Mathematica)(Wolfram [1988]) that the orbit invariants described in Deift et al., [1986] are conserved up to order p using a *pth* order Runge-Kutta map to integrate Hamilton equations (2.1, 2.2).

In Toda flows, stiffness dominates sooner or later depending on the level of accuracy required (Watkins [1982]). When this happens, implicit integration should be used. In our numerical experiments, we have not used implicit schemes since we are mostly concerned with the early dynamics of the flow, when stiffness has not set in yet. However, we did consider more sophisticated explicit integration schemes such as the Bulirsch-Stoer method (Stoer and Bulirsch [1983]). Yet it is clear, as we mentioned earlier, that schemes that numerically integrate the flow are not competitive algorithms for the algebraic eigenvalue problem at this point. For dense matrices for instance, each commutator evaluation scales as n^3. More precisely, the differential equation for a matrix element a_{ij} of a real symmetric matrix A is (for $j > i$)

$$\dot{a}_{ij} = a_{ij}(a_{jj} - a_{ii}) + 2\sum_{k>j} a_{ik}a_{kj} - 2\sum_{k<i} a_{ik}a_{kj}. \tag{2.3}$$

Six such commutator evaluations are needed for a single application of the explicit $5th$ order Runge-Kutta map.

In contrast with the computationally demanding integration of Toda flows, it is possible to reduce a symmetric dense matrix to tridiagonal form in $O(n^3)$ operations using Householder reflections. The resulting tridiagonal eigenvalue problem can then be solved, using the QR algorithm for example. Watkins [1982] has shown that the QR algorithm for symmetric tridiagonal matrices remains superior to the numerical integration of the flow, both for speed and accuracy. Although it is true that integration of Toda flows can greatly benefit from parallelism, as we shall see in the next section, conventional algorithms that solve the symmetric eigenvalue problem parallelize as well. For symmetric tridiagonals for example, parallel bisection or divide and conquer techniques have a high degree of parallelism, more so than QR. Householder reduction to tridagonal form can also be implemented very efficiently on parallel architectures (Lichtenstein and Johnsson [1993], CMSSL release notes [1993]).

In view of the above analysis, our goal was to visualize the flow in real time, rather than solving an eigenvalue problem per se. Although the matrix commutators in Hamilton equations (2.1, 2.2) can be evaluated in less operations than two matrix multiplies, we did not take advantage of this in order to benefit from the particularly efficient implementation of dense matrix multiplication on the Connection machine (Mathur and Johnsson [1991]). Of course, one should take advantage of the band preserving property of the flow in the evaluation of the commutators (2.1, 2.2) whenever possible. For tridiagonal matrices for instance, equation (2.3) reduces to

$$\begin{cases} \dot{a}_i &= 2(b_i^2 - b_{i-1}^2), \quad (b_0 = b_n = 0) \\ \dot{b}_i &= b_i(a_{i+1} - a_i) \end{cases}$$

where a_i and b_i are the diagonal and subdiagonal elements, respectively. Assuming the n pairs of matrix elements (a_i, b_i) to be distributed on a one dimensional array of n processors, the evaluation of the above expressions only requires communication between neighboring processors and local computation. Each commutator is therefore evaluated in $O(1)$ operations as opposed to $O(n)$ operations on a conventional, serial computer. It is clear that data locality, a very desirable feature for distributed memory machines, is inversely proportional to the bandwidth of the matrix.

Finally, we wish to say a few words regarding the nonsymmetric case. Upper Hessenberg matrices (i.e. matrices such that $a_{ij} = 0$ if $i - j \leq 2$) are to the nonsymmetric eigenvalue problem what symmetric tridiagonal matrices are to the symmetric case. The fixed point of the nonsymmetric flow is the real Schur form (Chu [1984]), a block upper triangular matrix. Each pair of complex conjugate eigenvalues contributes a block of size two by two. Interestingly the commutator equation (2.2) has a strong data locality for upper Hessenberg matrices and can be evaluated efficiently in parallel. Call h_i the leftmost subdiagonal elements (i.e. $h_i = a_{i+2,i}$ for $i = 1$ to $n - 2$), equation (2.3) becomes

$$a_{i,j}^{\cdot} = h_i a_{i+1,j} - h_{i-1} a_{i-1,j} + h_i a_{i,j+1} - h_{i-1} a_{i,j-1}.$$

All derivatives $a_{i,j}^{\cdot}$ can be evaluated concurrently as follows. First, each h_i is broadcast to its corresponding $(i + 2)th$ row and ith column elements. This can be done in $O(log(n))$ time complexity on parallel machines using parallel prefix operations. Next, each matrix element has to access the data from its four neighbors and perform some simple arithmetic. The communication pattern involved is a five-point stencil as shown below.

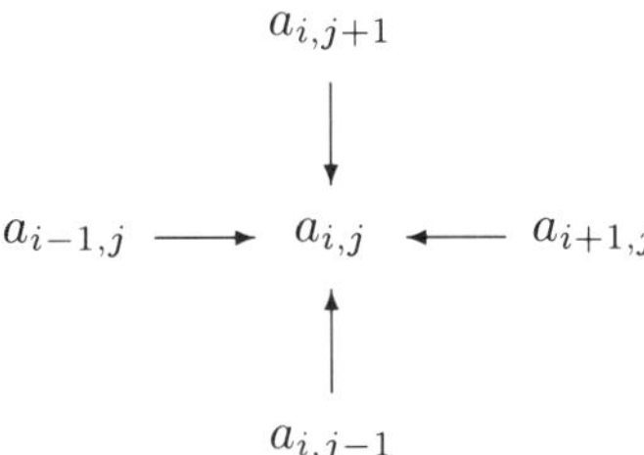

This communication pattern, which is that of a finite difference discretization of a two dimensional elliptic operator, is a very efficient operation on massively parallel architectures.

It is worth noting that in contrast to Toda flows, conventional algorithms for the nonsymmetric eigenvalue problem do not parallelize as easily as they do in the symmetric case.

3 Orbit visualization on the Connection Machine

Numerical experiments are performed on the Connection Machine 2 (CM2), a distributed memory architecture which consists of 65,536 bit-serial processors (for the largest configuration), with 16 such processors implemented on a single chip, the CM chip. Two chips with bit-serial processors share a floating-point unit. For this work, programs were developed in CMFortran, a data parallel language that implements the array-processing features of Fortran 90. In the case of matrix computation, each matrix is represented as an array and each matrix entry is owed by a processor. In matrix orbit computation, matrix entries evolve in time according to differential equations. Data parallelism in this context means that the coupled differential equations are all solved in unison, instead of being processed one after the other on a conventional, serial computer. Furthermore, the simultaneous update of all matrix entries allows for the real-time visualization of the flow. Following an integration step, the matrix elements are assigned a color value and the resulting image is transferred to a CM graphic display which consists of a framebuffer module and a high resolution color monitor. Using a regular red-green-blue color lookup table, the absolute values of the matrix elements are scaled and their magnitude converted to a 8-bit color value where red is attributed to the largest value and black to the smallest. Selected frames shown below are converted to gray scale.

In Figure 1 we show snapshots taken from the orbital evolution of a Toeplitz matrix of size 128. A matrix is said to be Toeplitz when the entries are constant along each diagonal. An identical Toeplitz matrix was used earlier to display the rich dynamics of the Jacobi algorithm on the Connection Machine (Edelman [1991]). For both Jacobi and isospectral flows, not all matrices produce interesting patterns to observe. Besides their practical interest, Toeplitz matrices seem to be good candidates for the visualization of matrix factorization dynamics. As we shall see, the Toeplitz property is not preserved by the flow. The flow is visualized in real time on a 8K processors CM2. The internal time at which the pictures were taken are indicated in the figure. At $t = 0$, the entries of the matrix are set to zero in the $2m + 1$ interior diagonals and to one elsewhere, with $m = 20$

$$a_{ij} = \begin{cases} 0 & for \ | \, j - i \, | < m \\ 1 & otherwise. \end{cases}$$

We are mostly interested in the transient dynamics. The matter of fact is that convergence to diagonal form is extremely slow in this case anyway because of the occurence of multiple zero eigenvalues. Although the matrix is symmetric with respect to the north-east, south-west diagonal, the spectrum is not symmetric with respect to the origin. Note that only the matrix entries of absolute magnitude less or equal to one are mapped onto the gray scale. All matrix elements larger than one in absolute value are white. From $t = 0$ to $t = 0.1$, we see a front wave developing outside the inner band from the lower right corner and moving toward the upper left corner. During the same period of time, another front wave is propagating in the opposite direction within the band. After $t = 0.1$, the outer matrix elements are close to zero and most of the action is concentrated inside the band. The regions of interest are the two (m, m) matrix blocks located at both extremes of the main diagonal. Particularly interesting are the patterns that develop in the lower right corner.

4 Abstract matrix factorization

The generalized Toda flow equations (2.1, 2.2) arise from the natural decomposition of the Lie algebra of the linear group into the algebras of lower triangular and skew symmetric matrices. Other Lie algebra decompositions give rise to group factorizations and associated matrix factorization algorithms such as the LU or Cholesky algorithms (Watkins [1984]).

As shown by Chu and Norris [1988], it is not necessary to use Lie algebra decompositions to construct dynamical systems that factorize matrices. Suppose A is a symmetric matrix of order n and choose a set of indices $\Delta = \{(i, j), 1 \leq j < i \leq n\}$. Define a pattern X with

$$x_{ij} = \begin{cases} a_{ij} & for (i, j) \in \Delta \\ 0 & otherwise. \end{cases}$$

The solution to the system

$$\dot{A} = [A, X - X^T] \tag{4.1}$$

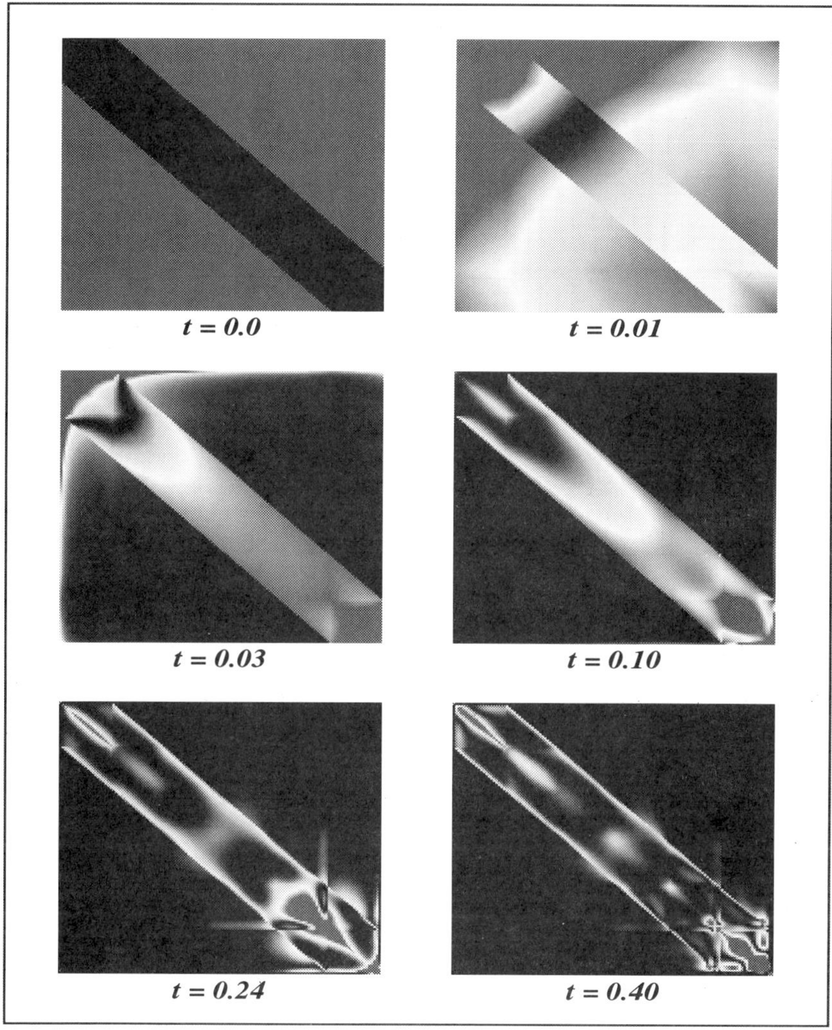

Figure 1 Isospectral deformations of a Toeplitz matrix of order 128. Absolute values of matrix entries in the interval $[0, 1]$ are mapped onto a gray scale. Entries larger then 1 appear in white. Time shown is internal time at which snapshots were taken.

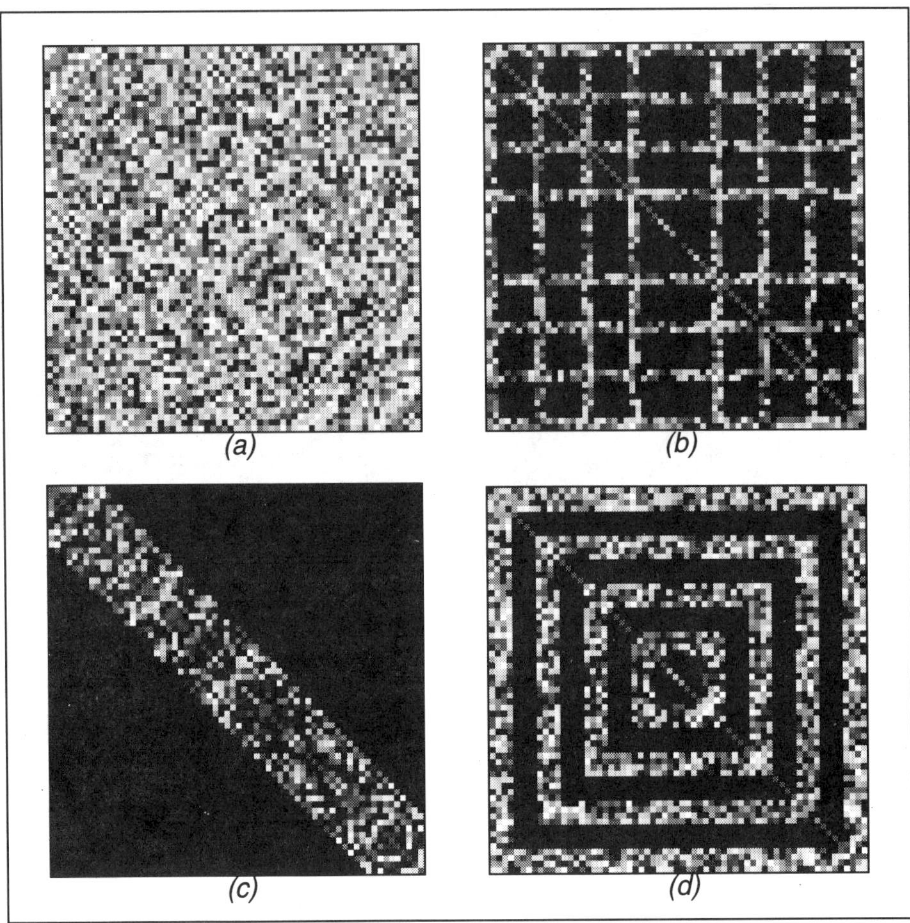

Figure 2 Abstract matrix factorizations of a random matrix with elements between 0 and 1. The matrix is shown in (a). Pictures (b), (c) and (d) are limit points of isospectral trajectories based at (a) but with different abstract patterns. Absolute values of matrix entries in the interval $[0,1]$ are mapped onto a gray scale.

is defined for all time and converges to a limit point as $t \to \infty$. Furthermore $a_{ij} \to 0$ if $(i,j) \in \Delta$. This is the generalized Schur decomposition theorem (Chu and Norris [1988]).

Given a matrix A, it is thus possible to construct a matrix $\tilde{A}$, similar to A (i.e. with the same spectrum), but with an arbitrary pattern of zero entries. For example, selecting Δ to be all matrix elements but those on the diagonal and the subdiagonal leads to a dynamical system which is a continuous realization of the Householder reduction to tridiagonal form. However, unlike the Householder case, the entries of the tridiagonal matrix are sorted in nondecreasing order. This is a salient feature of Toda flows which slows down the convergence drastically because of the occurence of saddle points (the tridiagonal matrices with a different ordering of the diagonal). It is certainly possible to enforce any ordering along the diagonal using gradient flows (Brockett [1991], Chu [1992]). However, as we mentioned earlier gradient flows do not preserve the bandwidth. This of course does not matter for initially dense matrices (which can be reordered anyway) but is a major obstacle for reducing a banded matrix to tridiagonal form for example. From a practical purpose, the matrix converges to diagonal form just about as quickly as it converges to tridiagonal form.

In Figure 2 we illustrate several abstract matrix factorizations. All the displayed matrices are similar. The initial condition for the flows is a random matrix with elements between 0 and 1 displayed in Figure 2a. The three factorizations shown in 2b, 2c and 2d are limit points of the flow generated by equation (4.1) with different patterns X. Only the matrix elements in $[0,1]$ are mapped to the gray scale. By the time the flow has reached its limit point, the matrix is highly diagonally dominant. Furthermore the diagonal elements are sorted in all cases. Except for the banded form, it is not clear that such factorizations would offer any interest in practice. Still it is amusing that similar matrices of arbitrary shape can be tailored using dynamical systems.

References

Brockett, R.W. [1991] *Dynamical systems that sort lists, diagonalize matrices, and solve linear programming problems*, Linear Algebra Appl., **146**, 79–91.

CMSSL release notes for the CM-200, Thinking Machines Corporation [1993], 164–194. *CMSSL for CM Fortran: CM-5 Edition*, Vol I, Thinking Machines Corporation [1993], 275–308. CMSSL is the Connection Machine Scientific Software Library, which is available on Connection Machine systems.

Chu, M.T. [1984] *The generalized Toda flow, the QR algorithm and center manifold theory*, SIAM J. Alg. Disc. Meth., **5**, 187–201.

_______ [1992] *Matrix differential equations: A continuous realization process for linear algebra problems*, Nonlinear Anal., TMA, **18**, 1125–1146.

Chu, M.T., and Norris, L.K. [1988] *Isospectral flows and abstract matrix factorizations*, SIAM J. Numer. Anal., **25**, 1383–1391.

Deift, P., Nanda, T., and Tomei, C. [1983] *Ordinary differential equations and the symmetric eigenvalue problem*, SIAM J. Numer. Anal., **20**, 1–22.

Deift, P., Li, L.C., Nanda, T., and Tomei, C. [1986] *The Toda flow on a generic orbit is integrable*, Comm. Pure Appl. Math., **39**, 183–232.

Deift, P., Li, L.C., and Tomei, C. [1989] *Matrix factorizations and integrable systems,* Comm. Pure Appl. Math., **42**, 443–521.

Edelman, A. [1991] *Snapshots of mobile Jacobi;* in Numerical Linear Algebra, Digital Signal Processing and Parallel Algorithms, NATO ASI Series F, Computer and system sciences, **70** (G. Golub and P. van Dooren, eds.) Springer Verlag, Berlin, 485–488.

Mathur, K.K., and Johnsson, S.L. [1991] *Multiplication of matrices of arbitrary shapes on a data parallel computer,* Thinking Machines Corporation Technical Report TR-216. Submitted to the Journal of Parallel Computing.

Lichtenstein, W., and Johnsson, S.L. [1993] *Block cyclic dense linear algebra,* Thinking Machines Corporation Technical Report TR-215, 1992. To be published in SIAM J. Sci. Comp., **14**.

Press, W.H., and Teukolsky, S.A. [1992] *Adaptive stepsize Runge-Kutta integration,* Computers in Physics, **6**, 188–191.

Stoer, J., and Bulirsch, R. [1983] *Introduction to Numerical Analysis.* Springer Verlag, Berlin, 458.

Symes, W.W. [1982] *The QR algorithm and scattering for the finite non-periodic Toda lattice,* Physica 4D, 275–280.

———[1980] *Hamiltonian group actions and integrable systems,* Physica 1D, 339–374.

Watkins, D.S. [1984] *Isospectral flows,* SIAM Rev., **26**, 379–391.

———[1982] *Experience with the Toda flow method of calculating eigenvalues,* Washington State University Technical Report.

Wolfram, S. [1988] *Mathematica,* Addison-Wesley.

Fields Institute Communications
Volume 3, 1994

A List of Matrix Flows with Applications

Moody T. Chu
Department of Mathematics
North Carolina State University
Raleigh, North Carolina, USA
27695-8205

Abstract. Many mathematical problems, such as existence questions, are studied by using an appropriate realization process, either iteratively or continuously. This article is a collection of differential equations that have been proposed as special continuous realization processes. In some cases, there are remarkable connections between smooth flows and discrete numerical algorithms. In other cases, the flow approach seems advantageous in tackling very difficult problems. The flow approach has potential applications ranging from new development of numerical algorithms to the theoretical solution of open problems. Various aspects of the recent development and applications of the flow approach are reviewed in this article.

1 Introduction

A realization process, in a broad sense, means any deductive procedure that we use to comprehend and solve problems. In mathematics, especially for existence questions, a realization process often appears in the form of an iterative procedure or a differential equation. For years researchers have taken great effort to describe, analyze, and modify realization processes. Nowadays the success of this investigation is especially evident in discrete numerical algorithms. On the other hand, the use of differential equations to issues in computational mathematics has been found recently to afford fundamental insights into the structure and behavior of existing discrete methods and, sometimes, to suggest new and improved numerical methods.

This paper reflects upon a number of interesting continuous realization processes that have been proposed in the literature. Adopted from dynamical system

---

1991 *Mathematics Subject Classification*. Primary 15A18; Secondary 34A05.

This research was supported in part by National Science Foundation under grants DMS-9006135 and DMS-9123448.

Supported in part by the Ministry of Colleges and Universities of Ontario and the Natural Sciences and Engineering Research Council of Canada while visiting The Fields Institute.

terminology, each continuous realization process is referred to as a flow. Although the rich theory of differential equations can often be put to use, the dynamics of many of the proposed differential systems are not completely understood. The true impact on numerical algorithms also needs to be investigated further. The intention of this paper is to compile a list of flows with a brief description of possible applications so as to stimulate further interest and advance in the flow approach for realizing a problem.

The basic idea of continuous realization methods is to connect two abstract problems by a mathematical bridge. Usually one of the abstract problems is a made-up problem whose solution is trivial while the other is the real problem whose solution is difficult to find. The bridge usually takes the form of an integral curve for a certain ordinary differential equation that describes how the problem data, including the answer to the problem, are transformed from the simple system to the more complicated system. This idea will become clear in the next section.

Obviously, the most important issue in the flow approach is the assurance that a bridge connecting the two abstract problems does exist. The construction of a bridge can be motivated in several different ways: sometimes an existing discrete numerical method may be extended directly into a continuous model (Keller [1978], Chu [1988]); sometimes a differential equation arises naturally from a certain physical principles (Symes [1982], Watson [1986]); more often a vector field is constructed with a specific task in mind (Brockett [1991], Chu and Driessel [1990], Chu [1991], Chu and Driessel [1991]). We shall report the material only descriptively. For more extensive discussion, readers should refer to the bibliography.

We present the flows on a case-by-case basis. For brevity, we encapsulate the circumstances under which the discussion is set forth by the following labels:

ORIGINAL PROBLEM: The underlying problem that is to be solved.
DISCRETE METHOD: Basic schemes of any existing discrete methods.
MOTIVATION: Motivation or idea for the construction of a bridge (flow).
FLOW: The description of the differential equation.
INITIAL CONDITIONS: The starting point of the flow (The simple system).
SPECIAL FEATURES: Special features of the flow approach.
EXAMPLE: Examples or applications.
GENERALIZATION: Possible generalizations or new numerical schemes.

2   List of flows

2.1   Linear stationary flows

ORIGINAL PROBLEM: Solve the linear equation

$$Ax = b \qquad\qquad (2.1)$$

where $A \in R^{n \times n}$ and $x, b \in R^n$.

DISCRETE METHOD: Most linear stationary methods assume the form (Hageman and Young [1981])

$$x_{k+1} = Gx_k + c, \ k = 0, 1, 2, \ldots \qquad\qquad (2.2)$$

where

$$G \;=\; I - Q^{-1}A$$
$$c \;=\; Q^{-1}b$$

and Q is a splitting matrix of A.

MOTIVATION: Think of (2.2) as one Euler step with unit step size applied to a linear differential system.

FLOW:

$$\frac{dx}{dt} = -Q^{-1}Ax + c. \tag{2.3}$$

INITIAL CONDITIONS: $x(0)$ can be any point in R^n.

SPECIAL FEATURES: For global convergence of (2.3), only the inequalities

$$\Re\lambda_i(G) < 1 \tag{2.4}$$

for all eigenvalues λ_i of G are needed, which is much weaker than the would-be condition for the convergence of (2.2).

GENERALIZATION: Solving (2.3) by a numerical method amounts to a new iterative scheme, including highly complicated multistep iterative schemes (Eirola and Nevanlinna [1988], Chu [1988]).

2.2 Homotopy flows ORIGINAL PROBLEM: Solve the nonlinear equation

$$f(x) = 0 \tag{2.5}$$

where $f : R^n \longrightarrow R^n$ is continuously differentiable.

DISCRETE METHOD: A classical method is the Newton method (Ortega and Rheinboldt [1970])

$$x_{k+1} = x_k - \alpha_k (f'(x_k))^{-1} f(x_k). \tag{2.6}$$

MOTIVATION: At least two ways to motivate the continuous flows:

1. Think of (2.6) as one Euler step with step size α_k applied to the differential equation (Keller [1978])

$$\frac{dx}{ds} = -(f'(x))^{-1} f(x). \tag{2.7}$$

2. Connect the system (2.5) to a trivial system by, for example,

$$H(x,t) = f(x) - tf(x_0) \tag{2.8}$$

   where x_0 is an arbitrarily fixed point in R^n (Allgower [1980]), (Allgower and Georg [1980]). Generically, the zero set $H^{-1}(0)$ is a one-dimensional smooth manifold.

FLOW: Either (2.7) or

$$\begin{bmatrix} f'(x) & -\frac{1}{t}f(x) \\ \frac{dx}{ds} & \frac{dt}{ds} \end{bmatrix} \begin{bmatrix} \frac{dx}{ds} \\ \frac{dt}{ds} \end{bmatrix} = \begin{bmatrix} 0 \\ 1 \end{bmatrix}. \tag{2.9}$$

INITIAL CONDITIONS: For (2.7), $x(0)$ can be arbitrary. For (2.9), $x(0) = x_0$ and $t(0) = 1$.

SPECIAL FEATURES:

1. The flow of (2.7) satisfies $f(x(s)) = e^{-s} f(x(0))$. That is, the flow moves in the direction along which $\|f(x)\|$ is exponentially reduced.

2. Properties of f and the selection of x_0 must be taken into account in (2.9) in order that the bridge really makes the desired connection to $t = 0$ (Allgower [1980], Allgower and Georg [1980]).

EXAMPLE: Successful applications with specially formulated homotopy functions include eigenvalue problems (Chu, Li and Sauer [1988], Li and Rhee [1989]), nonlinear programming problem (Garcia and Zangwill [1981]), physics applications and boundary value problems (Rheinboldt [1986], Watson [1986]), and polynomial systems (Li, Sauer and Yorke [1987], Morgan [1987]).

2.3 Scaled Toda flows ORIGINAL PROBLEM: Reduce a square matrix $A_0 \in R^{n \times n}$ to a certain canonical form (Horn and Johnson [1991]), e.g., triangularization, so as to solve the eigenvalue problem

$$A_0 x = \lambda x. \tag{2.10}$$

DISCRETE METHOD:

1. For triangularization, use the unshifted QR algorithm:

$$A_k = Q_k R_k \implies A_{k+1} = R_k Q_k \tag{2.11}$$

   where $Q_k R_k$ is the QR decomposition of A_k; or any other QR-type algorithms, e.g., the LU algorithm (Golub and Van Loan [1989], Wilkinson [1965]).

2. For a general non-zero pattern which A_0 is reduced to, no discrete method is available.

FLOW:

1. The Toda flow

$$\frac{dX}{dt} = [X, \Pi_0(X)] \tag{2.12}$$

   where $[A, B] = AB - BA$, $\Pi_0(X) = X^- - X^{-T}$ and X^- is the strictly lower triangular part of X.

2. The scaled Toda flow

$$\frac{dX}{dt} = [X, K \circ X] \tag{2.13}$$

   where K is a constant matrix and $\circ$ represent the Hadamard product.

INITIAL CONDITIONS: $X(0) = A_0$.
SPECIAL FEATURES:

1. The time-1 map of the Toda flow is equivalent to the QR algorithm (Deift, Nanda and Tomei [1983], Symes [1982]).

2. The time-1 map of the scaled Toda flow also enjoys a QR-like algorithm (Chu [1992a]).

3. For symmetric X, K is necessarily skew-symmetric. Asymptotic behavior of (2.13) is completely known (Chu [1992a]).

MOTIVATION: The Toda lattice originates as a description of a one-dimensional lattice of particles with exponential interaction. The connection between the Toda flow and the QR algorithm was discovered by Symes [1982].

EXAMPLE: Different choices of the scaling matrix K give rise to different isospectral flows, including many already proposed in the literature (Chu [1984], Chu [1992a], Della-Dora [1975], Watkins [1984], Watkins and Elsner [1988]). In particular, (2.13) can be used to generate special canonical forms that no other methods can (Chu and Norris [1988]).

2.4 Projected gradient flows ORIGINAL PROBLEM: Let $\mathcal{S}(n)$ and $\mathcal{O}(n)$ denote, respectively, the subspace of all symmetric matrices and the group of all orthogonal matrices in $R^{n \times n}$. Let $P(X)$ denote the projection of X onto a specified affine subspace of $\mathcal{S}(n)$. Then

$$\text{Minimize} \quad F(X) := \frac{1}{2} \|X - P(X)\|^2$$
$$\text{Subject to} \quad X \in \mathcal{M}(X_0) \tag{2.14}$$

where $\mathcal{M}(X_0) := \{X \in \mathcal{S}(n) | X = Q^T X_0 Q, Q \in \mathcal{O}(n)\}$ and $\| \cdot \|$ is the Frobenius norm.

DISCRETE METHOD: Depending upon the nature of the projection P, the problem may or may not have a discrete method (Chu and Driessel [1990]). The Jacobi method (Golub and Van Loan [1989]), for example, may be applied if $P(X) = \text{diag}(X)$.

FLOW:

$$\frac{dX}{dt} = [X, [X, P(X)]]. \tag{2.15}$$

MOTIVATION: The right hand side of (2.15) represents the negative of the gradient of F on the feasible set $\mathcal{M}(X_0)$.

INITIAL CONDITIONS: $X(0) = X_0$.

EXAMPLE:

1. The flow (2.15) may be employed to solve the least squares approximation problem subject to spectral constraints, the inverse eigenvalue problem and the eigenvalue problem (Chu and Driessel [1990]). For the latter, the flow (2.15) is a continuous analogue of the Jacobi method (Driessel [1986]).

2. The flow (2.15) generalizes Brockett's double bracket flow (Brockett [1991], Brockett [1989]) which, in turn, has been found to have other applications in sorting, linear programming and total least squares problems (Bloch, Brockett and Ratiu [1990], Bloch [1990]).

GENERALIZATION: The idea of projected gradient flow can be generalized to other types of approximation problems as will be seen below.

2.5 Simultaneous reduction flows ORIGINAL PROBLEM: Simultaneous reduction by two kinds of transformations:

1. Reduction by orthogonal similarity transformations: Given matrices $A_i \in R^{n \times n}$, $i = 1, \ldots, p$, and projection maps P_i onto specified subspaces,

$$\text{Minimize} \quad F(Q) := \frac{1}{2} \sum_{i=1}^{p} \|\alpha_i(Q)\|^2$$

$$\text{Subject to} \quad Q \in \mathcal{O}(n) \tag{2.16}$$

where $\alpha_i(Q) := Q^T A_i Q - P_i(Q^T A_i Q)$.

2. Reduction by orthogonal equivalence transformations: Given matrices $B_i \in R^{m \times n}$, $i = 1, \ldots, p$, and projection maps R_i,

$$\text{Minimize} \quad G(Q, Z) := \frac{1}{2} \sum_{i=1}^{p} \|\beta_i(Q, Z)\|^2$$

$$\text{Subject to} \quad Q \in \mathcal{O}(m) \tag{2.17}$$
$$Z \in \mathcal{O}(n)$$

where $\beta_i(Q, Z) := Q^T B_i Z - R_i(Q^T B_i Z)$.

DISCRETE METHOD: Very few theoretical results or even numerical methods are available for simultaneous reduction problems (Horn and Johnson [1991], Chu [1991]).

MOTIVATION: Compute the projected gradient of (2.16) and (2.17), respectively.

FLOW:

1. Orthogonal similar flow:

$$\frac{dX_i}{dt} = \left[X_i, \sum_{j=1}^{p} \frac{[X_j, P_j^T(X_j)] - [X_j, P_j^T(X_j)]^T}{2} \right]. \tag{2.18}$$

2. Orthogonal equivalence flow:

$$\frac{dY_i}{dt} = \sum_{j=1}^{p} \left\{ Y_i \frac{Y_j^T R_j(Y_j) - R_j^T(Y_j) Y_j}{2} + \frac{R_j(Y_j) Y_j^T - Y_j R_j^T(Y_j)}{2} Y_i \right\}. \tag{2.19}$$

INITIAL CONDITIONS: $X_i(0) = A_i$, $Y_i(0) = B_i$.

EXAMPLE: Here is a Jacobi flow for computing singular values of a single matrix:

$$\frac{dX}{dt} = X \frac{X^T \text{diag}(X) - (X^T \text{diag}(X))^T}{2} + \frac{\text{diag}(X) X^T - (\text{diag}(X) X^T)^T}{2} X. \tag{2.20}$$

GENERALIZATION: Two other related matrix flows (but not derived from projected gradient):

1. SVD flow:

$$\frac{dX}{dt} = X \Pi_0(X X^T) - \Pi_0(X^T X) X,$$
$$X(0) = A. \tag{2.21}$$

2. QZ flow:

$$\begin{aligned}
\frac{dX_1}{dt} &= X_1\Pi_0(X_2^{-1}X_1) - \Pi_0(X_1X_2^{-1})X_1, \\
\frac{dX_2}{dt} &= X_2\Pi_0(X_2^{-1}X_1) - \Pi_0(X_1X_2^{-1})X_2, \\
X_1(0) &= A_1, \\
X_2(0) &= A_2.
\end{aligned} \tag{2.22}$$

SPECIAL FEATURES:

1. The continuous realization processes (2.18) or (2.19) have the advantages that the desired form to which matrices are reduced can be almost arbitrary, and that if a desired form is not attainable then the limit point of the differential system gives a way of measuring the distance from the best reduced matrices to the nearest matrices that have the desired form.

2. Just as the Toda lattice (2.12) models the QR algorithm, the system (2.21) models the SVD algorithm (Chu [1986b]) for the $A \in R^{m \times n}$, and (2.22) models the QZ algorithm (Chu [1986a]) for the matrix pencil $(A_1, A_2) \in R^{n \times n} \times R^{n \times n}$.

2.6 Inverse eigenvalue flows ORIGINAL PROBLEM: Given a set of real numbers $\{\lambda_1, \ldots, \lambda_n\}$, consider two kinds of inverse eigenvalue problems:

1. Given $A_0, \ldots, A_n \in \mathcal{S}(n)$ that are mutually orthonormal, find $c = [c_1, \ldots, c_n]$ such that

$$A(c) := A_0 + \sum_{i=1}^{n} c_i A_i \tag{2.23}$$

has the prescribed set as its spectrum. The special case is where $A(c)$ is a Toeplitz matrix which is known, thus far, to be an open problem (Delsarte and Genin [1983]).

2. Find a symmetric non-negative matrix P that has the prescribed set as its spectrum.

DISCRETE METHOD: A few locally convergent Newton-like algorithms are available for the first problem (Friedland, Nocedal and Overton [1987], Laurie [1988]). Little is known for the non-negative matrix problem (Berman and Plemmons [1979]).

MOTIVATION: Minimize the distance between the isospectral surface and the set of matrices of desired form.

FLOW:

1. Inverse eigenvalue problem (Chu and Driessel [1990]):

$$\frac{dX}{dt} = [X, [X, A_0 + P(X)]] \tag{2.24}$$

where

$$P(X) = \sum_{i=1}^{n} < X, A_i > A_i, \tag{2.25}$$

and $< \cdot, \cdot >$ denotes the Frobenius inner product.

2. Inverse eigenvalue problem for non-negative matrices (Chu and Driessel [1991]):

$$\frac{dX}{dt} = [X, [X, Y]] \tag{2.26}$$

$$\frac{dY}{dt} = 4Y \circ (X - Y). \tag{2.27}$$

INITIAL CONDITIONS: For both (2.24) and (2.26), $X(0) = \mathrm{diag}\{\lambda_1, \ldots, \lambda_n\}$. For (2.27), $Y(0)$ can be any non-negative matrix.

GENERALIZATION:

1. For the inverse Toeplitz eigenvalue problem, the descent flow (2.24) may converge to a stationary point that is not Toeplitz. A new flow that seems to converge globally is (Driessel and Chu [1988]).

$$\frac{dX}{dt} = [X, k(X)] \tag{2.28}$$

where

$$k_{ij}(X) := \begin{cases} x_{i+1,j} - x_{i,j-1} & \text{if } 1 \leq i < j \leq n \\ 0 & \text{if } 1 \leq i = j \leq n \\ x_{i,j-1} - x_{i+1,j} & \text{if } 1 \leq j < i \leq n \end{cases} \tag{2.29}$$

2. The idea of (2.24) can be generalized to inverse singular value problem:

$$\frac{dX}{dt} = X \frac{X^T(B_0 + R(X)) - (B_0 + R(X))^T X}{2} \tag{2.30}$$

$$- \frac{X(B_0 + R(X))^T - (B_0 + R(X))^T X}{2} X \tag{2.31}$$

where

$$R(X) = \sum_{k=1}^{n} < X, B_k > B_k \tag{2.32}$$

and $B_0, B_1, \ldots, B_n \in R^{m \times n}$ are prescribed mutually orthonormal matrices. Recently insights drawn from (2.30) give rise to new iterative methods (Chu [1992b]).

2.7 Complex flows ORIGINAL PROBLEM: Most of the discussion hitherto can be generalized to the complex-valued cases. One such example is the nearest normal matrix problem (Higham [1989], Ruhe [1987]).

DISCRETE METHOD: The Jacobi algorithm (Ruhe [1987]) can be used.

MOTIVATION: The nearest normal matrix problem is equivalent to

$$\begin{aligned} \text{Minimize} \quad & H(U) := \frac{1}{2} \| U^* A U - \mathrm{diag}(U^* A U) \|^2 \\ \text{Subject to} \quad & U \in \mathcal{U}(n) \end{aligned} \tag{2.33}$$

where $\mathcal{U}(n)$ is the group of all unitary matrices in $C^{n \times n}$.

FLOW:

$$\frac{dU}{dt} = U \frac{[W, \mathrm{diag}(W^*)] - [W, \mathrm{diag}(W^*)]^*}{2}, \tag{2.34}$$

$$\frac{dW}{dt} = \left[W, \frac{[W, \mathrm{diag}(W^*)] - [W, \mathrm{diag}(W^*)]^*}{2}\right]. \tag{2.35}$$

INITIAL CONDITIONS: $U(0) = I$ and $W(0) = A$.

SPECIAL FEATURES: The putative nearest normal matrix to A is given by $Z := U(\infty)\mathrm{diag}(W(\infty))U(\infty)^*$ (Chu [1991]).

GENERALIZATION: Least square approximation by real normal matrices can also be done by a method described by Chu [1991]

$$\frac{dX}{dt} = \left[X, \frac{[X, A^T] - [X, A^T]^T}{2}\right]. \tag{2.36}$$

3 Conclusion

Most matrix differential equations by nature are complicated, since the components are coupled into nonlinear terms. Nonetheless, as we have demonstrated, there have been substantial advances in understanding some of the dynamics. For the time being, the numerical implementation is still very primitive. But most important of all, we think there are many opportunities where new algorithms may be developed from the realization process. It is hoped that this paper has conveyed some values of this idea.

References

Allgower, E. [1980] *A survey of homotopy methods for smooth mappings*; in Numerical Solution of Nonlinear Equations, Lecture Notes in Math., **878**, 1–29.

Allgower, E., and Georg, K. [1980] *Simplicial and continuation methods for approximating fixed points and solutions to systems of equations*, SIAM Review, **22**, 28–85.

Berman, A., and Plemmons, R.J. [1979] *Non-negative Matrices in the Mathematical Sciences*, Academic Press, New York.

Bloch, A.M., Brockett, R.W., and Ratiu, T.S. [1990] *A new formulation of the generalized Toda lattice equations and their fixed point analysis via the momentum map*, Bull. Amer. Math. Soc. (N.S.) **23**, 477–485.

Bloch, A.M. [1990] *Steepest descent, linear programming, and Hamiltonian flows*, Contemporary Math., **114**, 77–88.

Brockett, R.W. [1991] *Dynamical systems that sort lists, diagonalize matrices and solve linear programming problems*; in Proceedings of the 27th IEEE Conference on Decision and Control, IEEE, (1988), 799–803, and Linear Alg. Appl., **146**, 79–91.

______ [1989] *Least squares matching problems*, Linear Alg. Appl., **122/123/124**, 761–777.

Chu, M.T. [1988] *On the continuous realization of iterative processes*, SIAM Review, **30**, 375–387.

Chu, M.T., Li, T.Y., and Sauer, T. [1988] *Homotopy method for general λ-matrix problems*, SIAM J. Matrix Anal. Appl., **9**, 528–536.

Chu, M.T., and Norris, L.K. [1988] *Isospectral flows and abstract matrix factorizations*, SIAM J. Numer. Anal., **25**, 1383–1391.

Chu, M.T., and Driessel, K.R. [1990] *The projected gradient method for least squares matrix approximations with spectral constraints*, SIAM J. Numer. Anal., **27**, 1050–1060.

Chu, M.T. [1984] *The generalized Toda flow, the QR algorithm and the center manifold theorem*, SIAM J. Alg. Disc. Meth., **5**, 187–210.

———[1986a] *A continuous approximation to the generalized Schur decomposition*, Linear Alg. Appl., **78**, 119–132.

———[1986b] *On a differential equation approach to the singular value decomposition of bi-diagonal matrices*, Linear Alg. Appl., **80**, 71–79.

———[1991] *A continuous Jacobi-like approach to the simultaneous reduction of real matrices*, Linear Alg. Appl., **147**, 75–96.

Chu, M.T., and Driessel, K.R. [1991] *Constructing symmetric non-negative matrices with prescribed eigenvalues by differential equations*, SIAM J. Math. Anal., **22**, 1372–1387.

Chu, M.T. [1991] *Least squares approximation by real normal matrices with specified spectrum*, SIAM J. Matrix. Appl., **12**, 115–127.

———[1992a] *Scaled Toda-like flows*, SIAM J. Numer. Anal., (to appear).

———[1992b] *Numerical methods for inverse singular value problems*, SIAM J. Numer. Anal., **29**, 885–903.

Deift, P., Nanda, T., and Tomei, C. [1983] *Differential equations for the symmetric eigenvalue problem*, SIAM J. Numer. Anal., **20**, 1–22.

Della-Dora, J. [1975] *Numerical Linear Algorithms and Group Theory*, Linear Alg. Appl., **10**, 267–283.

Delsarte, P., and Genin, Y. *Spectral properties of finite Toeplitz matrices*, Proceedings of the 1983 International Symposium of Mathematical Theory of Networks and Systems, Beer-Sheva, Israel, 194–213.

Driessel, K.R., and Chu, M.T. [1988] *Can real symmetric Toeplitz matrices have arbitrary real spectra?* Preprint.

Driessel, K.R. [1986] *On isospectral gradient flow — solving matrix eigenproblems using differential equations*, in Inverse Problems (J. R. Cannon and U Hornung, eds.), ISNM 77, Birkhauser, 69–91.

Eirola, T., and Nevanlinna, O. [1988] *What do multistep methods approximate?* Numer. Math., **53**, 559–569.

Friedland, S., Nocedal, J., and Overton, M.L. [1987] *The formulation and analysis of numerical methods for inverse eigenvalue problems*, SIAM J. Numer. Anal., **24**, 634–667.

Garcia, C.B., and Zangwill, W.I. [1981] *Pathways to Solutions, Fixed Points, a Equilibria*, Prentice-Hall, Englewood Cliffs, NJ.

Golub, G.H., and Van Loan, C.F. [1989] *Matrix Computations*, 2nd ed., Johns Hopkins, Baltimore.

Hageman, L.A., and Young, D.M. [1981] *Applied Iterative Methods*, Academic Press, New York.

Higham, N.J. [1989] *Matrix nearness problems and applications*, Applications of Matrix Theory (S. Barnett and M. J. C. Gover, eds.), Oxford University Press, 1–27.

Horn, R.A., and Johnson, C.A. [1991] *Matrix Analysis*, Cambridge University Press, Cambridge.

Keller, H.B. [1978] *Global homotopies and Newton methods*; in Recent Advances Numerical Analysis (C. de Boor and G. Golub, eds.), Academic Press, New York, 73–94.

Laurie, D.P. [1988] *A numerical approach to the inverse Toeplitz eigenproblem*, SIAM J. Sci. Stat. Comput., **9**, 401–405.

Li, T.Y., and Rhee, N.H. [1989] *Homotopy algorithm for symmetric eigenvalue problems*, Numer. Math., **55**, 265–280.

Li, T.Y., Sauer, T., and Yorke, J. [1987] *Numerical solution of a class of deficient polynomial systems*, SIAM J. Numer. Anal., **24**, 435–451.

Morgan, A.P. [1987] *Solving Polynomial Systems Using Continuation for Scientific and Engineering Problems*, Prentice-Hall, Englewood Cliff, NJ.

Ortega, J.M., and Rheinboldt, W. [1970] *Iterative Solution of Nonlinear Equations in Several Variables*, Academic Press, New York.

Rheinboldt, W.C. [1986] *Numerical Analysis of Parameterized Nonlinear Equations*, John Wiley and Sons, New York.

Ruhe, A. [1987] *Closest normal matrix finally found!* BIT, **27**, 585–595.

Symes, W.W. [1982] *The QR algorithm and scattering for the finite non-periodic Toda lattice*, Physica, **4D**, 275–280.

Watkins, D.S. [1984] *Isospectral flows*, SIAM Rev., **26**, 379–391.

Watkins, D.S., and Elsner, L. [1988] *Self-similar flows*, Linear Alg. Appl., **110**, 213–242.

Watson, L.T., Billups, S.C., and Morgan, A.P. [1987] *HOMPACK: A suite of codes for globally convergent homotopy algorithms*, ACM Trans. Math. Software, **13**, 281–310.

Watson, L.T. [1986] *Numerical linear algebra aspects of globally convergent homotopy methods*, SIAM Review, **28**, 529–545.

Wilkinson, J.H. [1965] *The Algebraic Eigenvalue Problem*, Clarendon Press, Oxford.

Fields Institute Communications
Volume **3**, 1994

The Gibbs Variational Principle, Gradient Flows and Interior-Point Methods

Leonid Faybusovich
Department of Mathematics
University of Notre Dame
Notre Dame, Indiana, USA
46556-0398

Abstract. We analyze in detail relationships between the dynamical systems introduced in Faybusovich [1991d] and the Gibbs variational principle for a finite-state system of the statistical physics. The classical concept of duality is investigated in our context. In particular, the asymptotic behavior of the partition function is related to the solution of the dual linear programming problem.

1 Introduction

In the important paper (Brockett [1991]), R. Brockett introduced a class of dynamical systems and suggested using them for various purposes including for solving linear programming (LP) problems. Although Brockett's equations did not offer a practical way for solving LP problems (see Brockett and Wong [1991]), they indicated the existence of interior-point algorithms with the exponential rate of convergence. Bloch [1990, 1991] related Brockett's equations to Toda lattices in Moser's form and showed a close relationship between these equations and least-squares problems on Grassmann manifolds. He also described in a very elegant way the only known cases of explicit solutions to Brockett's equations (Bloch [1990]).

In our Harvard thesis, Faybusovich [1991d], we combined Brockett-Bloch ideas with some interior-point technique (for the review of interior-point methods of linear programming (see, for example, Goldfarb and Todd [1989] and Gonzaga [1992]). As a result a class of dynamical systems with interesting mathematical properties was obtained. In particular, the solutions of these systems converge to the optimal solutions of corresponding linear programming problems exponentially fast. We showed (Faybusovich [1991b]) that these systems are completely integrable, described their phase portraits and constructed action-angle variables. We also related

---

1991 *Mathematics Subject Classification.* 90C05, 82B05, 58F40.

Supported in part by the Ministry of Colleges and Universities of Ontario and the Natural Sciences and Engineering Research Council of Canada while visiting The Fields Institute.

these systems to toral actions on Kähler manifolds. The polyhedron defined by the constraints of linear programming problem arises as the image of the moment map corresponding to a toral action. The image under the moment map of the 'radial' part of the toral action gives our dynamical systems. The cases of nonlinear and time—varying cost function were addressed in Faybusovich [1991a] where also some topological properties of arising manifolds have been described.

In the present paper we consider in detail the relationship between our dynamical systems and the Gibbs variational principle of statistical physics. Understanding this relationship is extremely important for the construction of practical algorithms. In particular, we investigate the classical concept of duality in this context.

This paper is based on Faybusovich [1991d] (see also Faybusovich [1991c] and Faybusovich, (to appear)), and represents a fragment of our current efforts to construct discrete algorithms preserving nice properties of dynamical systems considered.

We feel that the work of Ammar and Martin [1986], Chu [1988], Deift, Nanda and Tomei [1983], and Watkins [1982] was very important for understanding the relationships between various discrete algorithms and continuous dynamical systems. See in this respect a recent comprehensive review paper (Brockett, (to appear)).

2 The maximization of the entropy on a polyhedron

Suppose that $e_1, ... e_n$ is a standard basis in R^n. We will denote by e the vector $e_1 + ... + e_n$ and $<,>$ will stand for the standard scalar product in R^n. Given $a_1, ... a_m \in R^n, b_1, ... b_m \in R$, consider the polyhedron $P \subset R^n$,

$$P = \{ p \in R^n : < a_i, p >= b_i, i = 1, 2, ... m, \tag{2.1}$$

$$< e, p >= 1, p = (p_1, ... p_n)^T, p_j \geq 0, j = 1, 2 \ldots n \}.$$

The optimization problem

$$F_c = < c, p > - \frac{p_1 \ln p_1 + \cdots + p_n \ln p_n}{\beta} \to \max, \tag{2.2}$$

$$p \in P, \tag{2.3}$$

is well-defined on P since the function $f(x) = x \ln x, x \in (0, \infty)$ admits a continuous extension $f(0) = 0$ to $[0, \infty)$. Here $\beta > 0$ is a parameter. The problem 2.2, 2.3 admits various interpretations. For example, it describes the Gibbs variational principle for the equilibrium distribution of a statistical physics system with a finite number of states. Under this interpretation $1/\beta$ should be treated as the temperature of the system and the cost function 2.2 represents the free energy of the system (Langford [1973]). We are looking for a solution of the problem 2.2, 2.3 in the form:

$$p_i(\beta, \lambda) = \mu_i(\beta, \lambda) / Z(\beta, \lambda), \tag{2.4}$$

$$\mu_i(\beta, \lambda) = \exp(\beta c_i + < \lambda, f_i >_m), i = 1, \cdots, m,$$

$$Z(\beta, \lambda) = \sum_{i=1}^{n} \mu_i(\beta, \lambda). \tag{2.5}$$

Here $\lambda \in R^m, f_i = (< a_1, e_i >, < a_2, e_i >, \cdots, < a_m, e_i >)^T$ and $<,>_m$ is the standard scalar product in R^m.

Remark 2.1 In statistical physics the sum $Z(\beta, \lambda)$ is usually called the partition function of the system described by the variational principle 2.2, 2.3.

Proposition 2.2 *Suppose that $\lambda \in R^m$ is such that $< a_i, p(\beta, \lambda) >= b_i, i = 1, \cdots, m$. Then 2.4 is a unique solution to the problem 2.2, 2.3.*

Proof For any $p \in P$ set $\delta = F_c(\beta, p(\beta, \lambda)) - F_c(\beta, p)$. Denote $p_i(\beta, \lambda)$ by ν_i. We have $\delta =< c, \nu_i - p_i > -(\nu_1 \ln \nu_1 + \ldots + \nu_n \ln \nu_n - p_1 \ln p_1 - \ldots - p_n \ln p_n)/\beta$. Further, $p_i \ln p_i - \nu_i \ln \nu_i = p_i \ln(p_i/\nu_i) + (p_i - \nu_i) \ln \nu_i$. Taking into account 2.4, 2.5, we obtain $\ln \nu_i = (\beta c_i + < \lambda, f_i >_m) - \ln Z(\beta, \lambda)$ and consequently $\delta = (p_1 \ln(p_1/\nu_1) + \ldots + p_n \ln(p_n/\nu_n))/\beta$. Using elementary inequality $f(x) = x \ln x \geq x - 1, x \geq 0, f(x) = x - 1$ only if $x = 1$ we obtain $\delta \geq < e, p - \nu > /\beta = 0$ and $\delta = 0$ if and only if $p_i = \nu_i, i = 1, \cdots, n$. $\square$

Corollary 2.3 *Suppose that $p(\beta, \lambda)$ is the optimal solution to 2.2 and 2.3 given by 2.4 and 2.5. Then*

$$F_c(\beta, p(\beta, \lambda)) = (\ln Z(\beta, \lambda) - < \lambda, b >_m)/\beta, \tag{2.6}$$

where $b = (b_1, \cdots, b_m) \in R^m$.

Our next goal is to describe the set $B \subset R^m$ consisting of those $b \in R^m$ for which the system of equations

$$< a_i, p(\beta, \lambda) >= b_i, i = 1, \cdots, m \tag{2.7}$$

is solvable. Here $p(\beta, \lambda)$ is described in 2.4 and 2.5.

Lemma 2.4 *Suppose that $g_i \in R^m, q_i > 0, i = 1, \cdots, n$. The function $Y : R^m \to R, Y(\lambda) = \ln(X(\lambda))$,*

$$X(\lambda) = \sum_{i=1}^{n} q_i \exp(< \lambda, g_i >)$$

is convex on R^m. Moreover, $D^2 Y(\lambda)(\xi, \xi) \geq 0$ and $D^2 Y(\lambda)(\xi, \xi) = 0$ if and only if

$$DY(\lambda)(\xi) =< \xi, g_1 >= \ldots =< \xi, g_n >, \xi \in R^m. \tag{2.8}$$

Remark 2.5 Here $D^2 Y(\lambda)$ stands for the Hessian of Y at the point $\lambda \in R^m$; $D(\lambda)$ is the Frechet derivative of the function Y at a point λ.

Proof We clearly have $D(\lambda)(\xi) =$

$$[q_1 < \xi, g_1 > \exp(< \lambda, g_1 >) + \ldots + q_n < \xi, g_n > \exp(< \lambda, g_n >)]/X(\lambda).$$

Further, $D^2 Y(\lambda)(\xi, \xi) =$

$$[(< \xi, g_1 > -DY(\lambda)(\xi))^2 q_1 \exp(< \lambda, g_1 >)+$$

$$\ldots + (< \xi, g_n > -DY(\lambda)(\xi))^2 q_n \exp((< \lambda, g_n >)]/X(\lambda).$$

Thus $D^2 Y(\lambda)(\xi, \xi) \geq 0$ and $D^2 Y(\lambda) = 0$ if and only if 2.8 holds. $\square$

Corollary 2.6 *The function $Y : (\beta, \lambda) \to \ln Z(\beta, \lambda)$ (see 2.4 , 2.5) is strictly convex on R^{m+1} provided the vectors $c, a_1, \cdots, a_m, e$ are linearly independent.*

Lemma 2.7 *Given $q_1, \cdots, q_n > 0, g_1, \cdots, g_n, b \in R^m$, consider the function $f(\lambda) =$*

$$\ln(q_1 e^{<\lambda, g_1>} + \ldots + q_n e^{<\lambda, g_n>}) - <\lambda, b>,$$

$\lambda \in R^m$. If $b \in int(conv(g_1, \cdots, g_n))$, then $f(\lambda) \to +\infty$, when $\| \lambda \| \to +\infty$. Here $int(conv(g_1, \cdots, g_n))$ is the interior of the convex hull of vectors $g_1, \cdots, g_n$ and $\| \lambda \|$ stands for the usual Euclidean norm of λ.

Proof We have $f(\lambda) \geq \ln(\min\{q_i : i \in [1, n]\} \exp(\max\{< \lambda, g_i - b >: i \in [1, n]\})$. But $max\{< \lambda, g_i - b >: i \in [1, n]\} = \max\{< \lambda, \mu >: \mu \in conv(g_1 - b, \cdots, g_n - b)\}$. Since $conv(g_1 - b, \cdots, g_n - b)$ contains a ball of radius ϵ for some positive ϵ, we obtain $f(\lambda) \geq \ln(\min\{q_i : i \in [1, n]\}) + \epsilon \| \lambda \|$. $\square$

Remark 2.8 This Lemma is essentially due to Langford [1973].

Theorem 2.9 *Suppose that the vectors $a_1, \cdots, a_m, e = e_1 + \ldots + e_n$ are linearly independent. For any $b \in int(conv(f_1, \cdots, f_n))$ the problem 2.2, 2.3 has a unique solution. This solution is of the form 2.4, 2.5, where $\lambda = \lambda(\beta, b)$ is a unique solution of the dual problem*

$$f_{\beta, b} = \ln Z(\beta, \lambda) - < \lambda, b > \to \min, \tag{2.9}$$

$\lambda \in R^m$, which depend smoothly on β, b. The optimal value of the cost function 2.2 is given by 2.6 and is a strictly concave function of b for any $\beta > 0$.

Proof The function $f_{\beta, b}$ is strictly convex on R^m. Moreover, by Lemma 2.4 $f_{\beta, b} \to +\infty$ when $\| \lambda \| \to +\infty$. Thus 2.9 has a unique solution $\lambda(\beta, b)$ which is described by the set of equations:

$$\nabla f_{\beta, b}(\lambda) = 0. \tag{2.10}$$

The solution to 2.10 depends smoothly on β, b by the implicit function theorem. Now 2.10 coincides with 2.7. By proposition 2.2 the problem 2.2 , 2.3 has a unique solution in the form 2.4, 2.5. The optimal value of the cost function is given by 2.6. It is a strictly concave function of b since it is obtained as a Legendre transform (Langford [1973]) of the strictly convex function $\lambda \to \ln Z(\beta, \lambda)$. $\square$

3 Dynamical systems that solve linear programming problems

We now extend the analysis of the previous section. Our goal is to find differential equations in β with solutions given by 2.4, 2.5. We need some elementary properties of polytopes. Any polytope is a convex hull of its extreme points (Brøndsted [1983]). In the case where a polytope P is described by 2.1, there is a well-known description of the set $E(P)$ of its extreme points (see e.g., Brøndsted [1983]). Given $p \in R^n$, denote by $J(p)$ the set $\{i \in [1, n] : p_i = 0\}$.

Proposition 3.1 *A point $p \in P$ is an extreme point of P if and only if the equalities $x_i = 0, i \in J(p), < a_i, x >= 0, i = 1, \cdots, m, < e, x >= 0$ imply $x = 0$.*

We call a polyhedron P described by 2.1 simple if for any $p \in E(P), | J(p) |= n - m - 1$. In other words, if the vectors $e_i, i \in J(p), a_j, j = 1, \cdots, m, e$ form a basis in R^n for any $p \in E(P)$. An extreme point $p \in E(P)$ is called nondegenerate if $| J(p) |= n - m - 1$. Denote by $int(P)$ the set $\{p \in P : p_i > 0, i = 1, \cdots, n\}$. Let, further, $T = span(a_1, \cdots, a_m, e)^{\perp}, \pi : R^n \to T$ be the orthogonal projection.

Proposition 3.2 *Suppose that $int(P) \neq \emptyset$ and $a_1, \cdots, a_m, e$ are linearly independent. Then $b \in T_n = int(conv(f_1, \cdots, f_n))$.*

Proof Let $V = \{p \in R^n\ :< e, p >= 0\}$. Consider the map $\eta : V \to R^m, \eta(\xi) = (< a_1, \xi >, \cdots, < a_m, \xi >)$. Since $a_1, \cdots, a_m, e$ are linearly independent, η is surjective. In particular, the image of any open ball in V is an open subset in R^m. Let $p \in int(P)$. There exists an $\epsilon > 0$ such that $p + \xi \in \{p \in R^n\ :< e, p >= 1, p_i \geq 0, i = 1, \cdots, n\}$ for any $\xi \in V$ such that $\| \ \xi \ \| < \epsilon$. But then the set $\{b + \eta(\xi) : \xi \in V, \| \ \xi \ \| < \epsilon\} \subset conv(f_1, \cdots, f_n)$. $\square$

Proposition 3.3 *Under assumptions of Proposition 3.2 the map*

$$\varphi : int(P) \to T, \varphi(p) = \pi((\ln p_1, \cdots, \ln p_n)^T)$$

is a diffeomorphism of $int(P)$ onto T.

Proof Suppose $c \in T$. Consider the corresponding problem 2.2 and 2.3. We know from Theorem 2.9 and Proposition 3.2 that this problem has a unique solution $p(\beta, \lambda)$ as described by 2.4, 2.5. We clearly have $\varphi(p(\beta, \lambda)) = \beta c, p(\beta, \lambda) \in int(P)$. Thus φ is surjective. If $\varphi(p) = \varphi(p'), p, p' \in int(P)$, then $f = (\ln(p_1/p_1'), \cdots, \ln(p_n/p_n'))^T = \lambda_1 a_1 + \ldots + \lambda_m a_m + \mu e$ for some real $\lambda_1, \cdots, \lambda_m, \mu$. Denote $< f, e_i >$ by γ_i and let $p_i(t) = \exp(\gamma_i t) p_i', i = 1, \cdots, n, t \in R$. It is clear that $p = p(1)$. Besides,

$$\frac{d}{dt} < f, p(t) >= \sum_{i=1}^{n} \gamma_i p_i(t) < f, e_i >=$$

$$\sum_{i=1}^{n} \gamma_i^2 p_i(t) > 0$$

unless $f = 0$. This means either $p = p'$ or $< f, p >>< f, p' >$. But $< f, p >=< f, p' >= \lambda_1 b_1 + \cdots + \lambda_m b_m + \mu$ since $p, p' \in P$. Hence $p = p'$, i.e. φ is an injection. Finally, if $\xi \in R^n, p \in int(P)$, we have

$$< \nabla\varphi(p), \xi >= \pi((\xi_1/p_1, \cdots, \xi_n/p_n)^T).$$

Now conditions $< \nabla\varphi, \xi >= 0, < a_i, \xi >= 0, i = 1, \cdots, m, < e, \xi >= 0$ easily imply

$$\sum_{i=1}^{n} \xi_i^2/p_i = 0$$

or $\xi = 0$. Thus $D\varphi(p)$ is an isomorphism of $\{\xi \in R^n\ :< a_i, \xi >= 0, i = 1, \cdots, m, < e, \xi >= 0\}$ onto T. By the implicit function theorem φ is an isomorphism of $int(P)$ onto T. $\square$

Remark 3.4 For an alternative proof of Proposition 3.3 under more restrictive assumptions see Faybusovich [1991b].

We denote by D(p) the diagonal matrix $diag(p_1, \cdots, p_n)$.

Lemma 3.5 *Suppose that the polytope P described by 2.1 is simple. Then for any $p \in P$ the vectors $D(p)a_1, \cdots, D(p)a_m, D(p)e$ are linearly independent.*

Proof Let $D(p)(\lambda_1 a_1 + \ldots + \lambda_m a_m + \lambda_{m+1} e) = 0$ for some $\lambda_i \in R$. This means that $< \lambda_1 a_1 + \ldots + \lambda_m a_m + \lambda_{m+1} e, e_i >= 0$ for any i such that $p_i > 0$. In other words, $\delta = \lambda_1 a_1 + \ldots + \lambda_m a_m + \lambda_{m+1} e \in span(e_i : i \in J(p))$. But for any $p \in P$ there exists an extreme point $p^* \in E(P)$ such that $J(p) \subset J(p^*)$ (see, for example, Brøndsted [1983]). This implies that $\delta \in span(e_i : i \in J(p^*))$, i.e. $\lambda_1 = \ldots = \lambda_{m+1} = 0$ since P is simple. $\square$

Introduce a $(m+1) \times n$ matrix $A = [a_1, \cdots, a_m, e]^T$. We have

Corollary 3.6 *If P is simple, then the $(m+1) \times (m+1)$ matrix $AD(p)A^T$ is invertible everywhere on P.*

We are now in position to introduce a smooth vector field which is defined on some open neighbourhood of P (see Faybusovich [1991a], [1991b], [1991d]) of a given simple polyhedron P. Namely, given $c \in R^n$, set

$$W_c(p) = D(p)c - D(p)A^T(AD(p)A^T)^{-1}AD(p)c, \tag{3.1}$$

$p \in U$,where $U = \{p \in R^n : A^T D(p)A$ is an invertible matrix $\}$.

Remark 3.7 Observe that $AW_c(p) = 0$ for any $p \in P$.

We want to describe solutions of 3.1 in terms of some variational problems. Consider the function $f : R^n \times R^m \times R \to R$,

$$f(p, \lambda, \beta) = ln(\sum_{i=1}^{n} \exp(\beta c_i + \sum_{j=1}^{m} a_{ji}\lambda_j)p_i) - \tag{3.2}$$

$$- \sum_{j=1}^{m} \lambda_j b_j.$$

Here $a_j = (a_{j1}, \cdots, a_{jn})^T j = 1, \cdots, m.$

Theorem 3.8 *For any $p \in P$, $\beta \in R$ there exists a unique solution $\lambda(p, \beta)$ of the problem $f(p, \lambda, \beta) \to \min, \lambda \in R^m$. Moreover,*

$$p_i(\beta) = \frac{\exp(\beta c_i + \sum_{j=1}^{m} a_{ji}\lambda_j(p, \beta))p_i}{Z(p, \beta)}, \tag{3.3}$$

$i = 1, \cdots, n,$

$$Z(p, \beta) = \sum_{i=1}^{n} \exp(\beta c_i + \sum_{j=1}^{m} a_{ji}\lambda_j(p, \beta))p_i \tag{3.4}$$

is the solution to 3.1 such that $p(0) = p$. The function $\lambda(p, \beta)$ smoothly depends on p, β and is a unique solution of the system of nonlinear equations:

$$< a_j, \nu(p, \lambda, \beta) >= b_j, j = 1, \cdots, m, \tag{3.5}$$

$$\nu_i(p, \lambda, \beta) = \mu_i(p, \lambda, \beta)/(\mu_1(p, \lambda, \beta) + \ldots + \mu_n(p, \lambda, \beta)), \tag{3.6}$$

$$\mu_i(p, \lambda, \beta) = \exp(\beta c_i + \sum_{j=1}^{m} a_{ij}\lambda_j)p_i, i = 1, \cdots, n.$$

Proof The properties of the solution $\lambda(p, \beta)$ of the problem $f(p, \lambda, \beta) \to \min$ are proved exactly as Theorem 2.9. We have to prove that 3.3 and 3.4 give a solution $p(\beta)$ of 3.1 such that p(0)=p. Consider, first, the case where $p \in int(P)$. Observe that $p(\beta) \in int(P)$ for any $p \in int(P), \beta \in R$. Now $\varphi(p(\beta)) = \beta\pi(c) + \varphi(p)$ (see Proposition 3.3). On the other hand, by 3.1

$$\varphi'(p(\beta))W_c(p(\beta)) = \pi(c). \tag{3.7}$$

Indeed, by the definition of $A, ImA^T = span(a_1, \cdots, a_m, e)$. From 3.1 it follows that

$$\varphi'(p(\beta)) \cdot W_c(p(\beta)) = \pi(c) - \pi(A^T(AD(p)A^T)^{-1}AD(p)c) = \pi(c),$$

since $\pi(span(a_1, \cdots, a_m, e)) = 0$. Thus we have

$$\frac{d}{d\beta}\varphi(p(\beta)) = \varphi'(p(\beta))p'(\beta) = \varphi'(p(\beta)) \cdot W_c(p(\beta)).$$

Since by Proposition 3.3 , $\varphi'(p)$ is injective for any $p \in P$, we obtain

$$p'(\beta) = W_c(p(\beta)), \beta \in R.$$

If p belongs to some face F of P, then $p(\beta) \in F$ for any β. We can prove that $p'(\beta) = W_c(p(\beta))$ applying the same reasoning as above to F instead of P. $\square$

From now on we denote $p_1 \ln p_1 + \ldots + p_n \ln p_n$ by $S(p)$.

Corollary 3.9 *Suppose that the polytope P has a nonempty interior and $p(\beta)$ is the solution to 2.2 and 2.3, $\beta > 0$. Then*

$$dp(\beta)/d\beta = W_{\pi c}(p(\beta)), \tag{3.8}$$

$$\lim p(\beta) = p_0, \beta \to 0,$$

where p_0 is a unique solution to the problem

$$-S(p) \to \max, p \in P.$$

Proof Indeed, in the proof of theorem 3.8 we have shown that 3.8 holds and that $\varphi(p(\beta)) = \beta\pi c$. Since φ is an isomorphism, it is clear that $\lim p(\beta), \beta \to 0$, exists and is equal to $\varphi^{-1}(0)$. Since $\pi\nabla S(\varphi^{-1}(0) = 0$, we have $\varphi^{-1}(0) = p_0$. $\square$

Remark 3.10 if $e/n \in P$ then $p_0 = e/n$, since $\varphi(e/n) = 0$.

Lemma 3.11 *Let $p \in int(P), \pi(c) \neq 0$. Then*

$$< c, W_c(p) >> 0.$$

Proof We have

$$W_c(p) = D(p)(c - A^T\Lambda(p)),$$

where

$$\Lambda(p) = (AD(p)A^T)^{-1}AD(p)c.$$

We know that $AW_c = 0$. Thus

$$< c, W_c(p) >=< c - A^T\Lambda(p), W_c(p) >=< D(p)^{-1}W_c(p), W_c(p) >\geq 0.$$

Now $< c, W_c(p) >= 0$ if and only if $W_c(p) = 0$ or $c - A^T \Lambda(p) = 0$, since $p \in int(P)$. Since $\pi(A^T \Lambda(p)) = 0$, we obtain $\pi(c) = 0$. A contradiction. Hence $< c, W_c(p) >>$ 0. $\square$

Theorem 3.12 *Suppose that $int(P) \neq \emptyset$ and $p(\beta))$ is the solution to 2.2 and 2.3, $\pi c \neq 0, \beta > 0.$, Then*

$$\frac{d}{d\beta} < c, p(\beta) >> 0, \frac{d}{d\beta} F_c(\beta, p(\beta)) < 0, \frac{dS}{d\beta}(p(\beta)) > 0, \frac{d}{d\beta}(S(p(\beta))/\beta) > 0.$$

Proof We have:

$$\frac{d}{d\beta} F_c(\beta, p(\beta)) = < \nabla F_c(\beta, p(\beta)), \dot{p}(\beta) > + \frac{\partial F_c}{\partial \beta}(\beta, p(\beta)).$$

Since $p(\beta)$ is the solution to 2.2, 2.3, we have $\nabla F_c(\beta, p(\beta)) \in span(a_1, \cdots, a_m, e)$. On the other hand, $\dot{p}(\beta) \in T = span(a_1, \cdots, a_m, e)^\perp$. Hence,

$$\delta = < \nabla F_c(\beta, p(\beta)), \dot{p}(\beta) >= 0.$$

This implies $dF_c(\beta, p(\beta))/d\beta = S((p(\beta))/\beta^2 < 0$, because $S(p) < 0$ for $p \in int(P)$. Further, $\delta = < c, \dot{p}(\beta) > - < \nabla S(p(\beta)), \dot{p}(\beta) > /\beta = 0$. By Lemma 3.11 $<$ $c, \dot{p}(\beta) >> 0$. Consequently, $dS(p(\beta))/d\beta > 0$. Finally,

$$\frac{d}{d\beta}(S(p(\beta)/\beta) = \frac{1}{\beta}\frac{d}{d\beta}S(p(\beta)) - \frac{1}{\beta^2}S(p(\beta)) > 0$$

for $\beta > 0$. $\square$

Corollary 3.13 *Let $M = \max\{< c, p >: p \in P\}$. Then (under assumptions of Theorem 3.12) $< c, p(\beta) >\to M, \beta \to +\infty$. For any positive ϵ we have*

$$M - < c, p(\beta) >< \epsilon$$

provided $\beta > \ln n/\epsilon$.

Proof Let $p^* \in P$ be such that $M = < c, p^* >$. We have $< c, p^* >\leq F_c(p^*, \beta)$ for any positive β, since $S(p) \leq 0, p \in P$. Moreover,

$$< c, p(\beta) >\leq M \leq F_c(p^*, \beta) \leq F_c(p(\beta), \beta). \tag{3.9}$$

By Theorem 3.12, we know that $\lim < c, p(\beta) >, \lim F_c(p(\beta), \beta), \beta \to +\infty$ exist and are equal. We conclude that $\lim < c, p(\beta) >= M, \beta \to +\infty$. We also have from 3.9 that

$$M - < c, p(\beta) >\leq -\frac{S(p(\beta))}{\beta}.$$

By Theorem 3.12, if $-S(p(\beta^*))/\beta^* \leq \epsilon$ for some $\beta^* > 0$ then this holds for any $\beta \geq \beta^*$. But $-S(p(\beta^*)) \leq -S(p_0) \leq \max\{-S(p) : p_1 + \ldots + p_n = 1, p_i \geq 0\} = \ln n$. Thus if we take $\beta^* = \ln n/\epsilon$, we obtain $-S(\beta^*)/\beta^* \leq \epsilon$. $\square$

4 Duality

Our next goal is to understand how the classical concept of duality in linear programming relates to our approach. First, we briefly recall some elementary properties of linear programming problems. Consider a linear programming problem

$$< c, p > \to \max, p \in P, \tag{4.1}$$

where P is described in 2.1. Suppose that P is simple. Let $p^* \in E(P)$. Then the vectors $a_j, j = 1, \cdots, m, e, e_i, i \in J(p^*)$, form a basis in R^n. In particular, there exist $h_i, i \in J(p^*)$, such that

$$< e, h_i > = < a_j, h_i > = 0, < e_{i'}, h_i > = \delta_{i', i}, \tag{4.2}$$

$j \in [1, m], i, i' \in J(p^*)$. Let

$$c = \sum_{j=1}^{m} \lambda_j a_j + \mu e + \sum_{i \in J(p^*)} \Delta_i e_i. \tag{4.3}$$

We clearly have $\Delta_i = < c, h_i >, i \in J(p^*)$.

Proposition 4.1 *An extreme point $p^* \in E(P)$ is a solution to 4.1 if and only if $\Delta_i \leq 0, i \in J(p^*)$.*

For a proof see, for example, Brøndsted [1983]. Given $J \subset [1, n]$, we define the face $F(J)$ of P as $\{p \in P : p_i > 0, i \notin J, p_j = 0, j \in J\}$. The family $F(J), J \subset [1, n]$ defines a decomposition of P with the following properties:

$$P = \bigcup_{J \subset [1, n]} F(J), F(J) \bigcap F(J') = \emptyset$$

if $J \neq J'$. The sets $F(J)$ are called (open) faces of P. In particular, $F(\emptyset) = int(P)$ and if $p \in E(P)$, then $F(J(p)) = \{p\}$. For any nonempty face $F(J)$ denote by $\dim F(J)$ the number $n - (m + 1) - cardJ$. Suppose that the function $p \to < c, p >$ is not constant on any one-dimensional face of P (one-dimensional faces are called edges). Then the problems

$$< c, p > \to \max(\min), p \in Cl(F(J)),$$

have unique solutions $p_{\max}(J), p_{\min}(J)$. Here we denoted by $Cl(F(J))$ the closure of $F(J)$ in the standard topology in R^n. Observe that $p_{\max}(J), p_{\min}(J)$ are extreme points of P. More generally, $E(Cl(F(J)) \subset E(P)$ for any face $F(J)$ of P.

Theorem 4.2 *Suppose that $< c, p >$ is not constant on any edge of P. Then any face $F(J)$ is an invariant manifold for the vector field W_c. If $p(t)$ is an integral curve of W_c such that $p(0) \in F(J)$, then*

$$\lim p(t) = p_{\max}(J), t \to +\infty, \lim p(t) = p_{\min}(J), t \to -\infty.$$

In particular, if $p(0) \in int(P)$, then $p(t)$ converges to the solution p^* of 4.1 when $t \to +\infty$. Moreover,

$$\mid p_i(t) \mid \leq K(p(0)) e^{\Delta_i t}, t \geq 0, i \in J(p^*), \tag{4.4}$$

where Δ_i are defined in 4.3. Here $K : P \to R$ is some positive function.

Remark 4.3 Thus solutions to dynamical system 3.1 converge to the solution p^* of the linear programming problem 4.1 exponentially fast.

For the proof see Faybusovich [1991b].

Corollary 4.4 *For any $p \in int(P)$*

$$\lim(\lambda_j(p,\beta))/\beta = -\lambda_j, j = 1, \cdots, m \tag{4.5}$$

$$\lim \ln[Z(p,\beta)]/\beta) = \mu, \tag{4.6}$$

$$\lim(f(p,\lambda(p,\beta))/\beta = \mu + \sum_{j=1}^{m} \lambda_j b_j, \tag{4.7}$$

$\beta \to \infty.$

Remark 4.5 Here $\lambda(p,\beta), Z(p,\beta), f(p,\lambda,\beta)$ are defined in 3.2, 3.3, 3.4 and λ_j, μ are defined in 4.3.

Proof Let $p^* \in E(P)$ be a solution to 4.1, $I = [1,n] \setminus J(p*)$. Let $p(\beta)$ be the integral curve of W_c such that $p(0) = p$. We know that $[\ln p_i(\beta)]/\beta \to 0, \beta \to +\infty, i \in I$ (since $p_i^* > 0$). By 3.2, 3.3 we have

$$[\ln p_i(\beta)]/\beta = c_i + \sum_{j=1}^{m} a_{ji}(\lambda_j(p,\beta)/\beta) - [\ln Z(\beta)]/\beta,$$

$i \in I$. On the other hand by 4.3

$$c_i + \sum_{j=1}^{m} a_{ji}(-\lambda_j) - \mu = 0, i \in I.$$

This yields

$$[\ln p_i(\beta)]/\beta = \sum_{j=1}^{m} a_{ji}(\lambda_j(p,\beta)/\beta + \lambda_j) + \mu - [\ln Z(\beta)]/\beta,$$

$i \in I$. Observe now that the system of linear equations

$$\sum_{j=1}^{m} a_{ji}x_j + x_{m+1} = 0, i \in I \tag{4.8}$$

has only zero solution. Indeed, 4.8 is equivalent to

$$x_1 a_1 + x_2 a_2 + \ldots + x_m a_m + x_{m+1}e \in span(e_k, k \in J(p*)).$$

Thus

$$\lambda_j(p,\beta)/\beta + \lambda_j = \sum_{i \in I} d_{ij}(\ln p_i(\beta)/\beta)),$$

$$\mu - [\ln Z(\beta)]/\beta = \sum_{i \in I} d_{i,m+1}(\ln p_i(\beta)/\beta)),$$

for some constants d_{ij}. Thus we obtain 4.5 and 4.6. Now 4.7 follows from 4.5, 4.6 and 3.2. $\quad \square$

Recall that $\pi : R^n \to T$ is the orthogonal projection of R^n onto orthogonal complement T of the $span(a_1, \cdots, a_m, e)$. Consider the following (primal-dual) optimization problem:

$$\psi_\beta(p, z, \beta) = < z, p > - S(p)/\beta - \frac{\sum_{i=1}^{n} e^{\beta z_i}}{\beta} \to \max, \tag{4.9}$$

$$< a_i, p >= b_i, i = 1, \cdots, m, \pi c = \pi z, \tag{4.10}$$

$$< e, p >= 1, p \geq 0, z \leq 0. \tag{4.11}$$

The next theorem paves the way to the construction of primal-dual algorithms.

Theorem 4.6 *Let* $p(\beta)$ *be the solution to* 2.1, 2.2, *and* 2.3, *be* $z_i(\beta) = [\ln p_i(\beta)]/\beta$. *Then* $(p(\beta), z(\beta))$ *is the solution to* 4.9, 4.10, 4.11.

Proof We have $\varphi(p(\beta)) = \beta\pi c$. Thus $\pi z(\beta) = \pi c$. Since $\psi_\beta(p, z)$ is strictly convex function of (z, p) it is sufficient to verify that

$$\nabla_p \psi_\beta(p(\beta), z(\beta)) \in span(a_1, \cdots, a_m, e)$$

$$\nabla_z \psi_\beta(p(\beta), z(\beta)) = 0.$$

But $\pi\nabla_p\psi_\beta(p(\beta), z(\beta)) =$

$$\pi z(\beta) - \pi\nabla S(p(\beta))/\beta = \pi(c - \pi\nabla S(p(\beta)))/\beta = 0.$$

Further,

$$\frac{\partial\psi_\beta(p(\beta), z(\beta))}{\partial z_i} = p_i(\beta) - e^{\beta z_i} = 0. \qquad \square$$

5 Vector fields W_c as gradient flows

Given a polyhedron P described by 2.1, consider the map Φ:

$$R^n \to R^n, \Phi(x_1, \cdots, x_n) = (x_1^2, \cdots, x_n^2).$$

Suppose that $M = \Phi^{-1}(P)$.

Proposition 5.1 *If P is simple , then M is a smooth manifold of the dimension $n - (m + 1)$. Moreover, the functions $< e, \Phi(x) >$, $f_i(x) =< a_i, \Phi(x) >$ are independent everywhere on M. In particular, the tangent space $T_x(M)$ of M at a point $x \in M$ is described as follows:*

$$T_x(M) = \{\xi \in R^n :< D(x)a_i, \xi >= 0, i = 1, \cdots, m, < D(x)e, \xi >= 0\}. \tag{5.1}$$

For a proof see Faybusovich [1991b]. Denote by $\Pi(x)$ the orthogonal projector of R^n onto $T_x(M), x \in M$. By 5.1 we know that $T_x(M)$ is the orthogonal complement of the vector space $span(D(x)a_i, i = 1, \cdots, m, D(x)e)$. Consider the vector field X_c on M which is defined as follows:

$$X_c(x) = \Pi(x)D(x)c. \tag{5.2}$$

Observe that X_c is simply the gradient of the function $f(x) =< c, \Phi(x) > /2$ restricted to M, provided M is endowed with the Euclidean metric g:

$$g(x; \xi, \eta) =< \xi, \eta >, \xi, \eta \in T_x(M).$$

Let $\Gamma = \{diag(\epsilon_1, \cdots, \epsilon_n) : \epsilon_i = \pm 1\}$. It is clear that Γ acts on M in a natural way.

Proposition 5.2 *The vector field X_c is invariant under the action of Γ, i.e.*

$$X_c(\alpha x) = \alpha X_c(x), \alpha \in \Gamma, x \in M.$$

Moreover,

$$T_x(\Phi)X_c(x) = 2W_c(\Phi(x)), x \in M.$$

In particular,if $x(t)$ is an integral curve of M, then $\Phi(x(2t))$ is an integral curve of W_c.

For a proof see Faybusovich [1991b]. Proposition 5.2 shows that there is a close relationship between vector fields X_c and W_c. In particular, one can easily describe the phase portrait of X_c. (See Faybusovich [1991b].) The interpretation of W_c as a gradient flow enables us to construct a Dikin-type algorithm for solving linear programming problems (Faybusovich, (to appear), Faybusovich [1991d]).

References

Ammar, G., and Martin, C. [1986] *The geometry of the matrix eigenvalue methods*, Acta App. Math., **5**, 239–278.

Bloch, A. [1991] *The Kähler structure of the total least squares problem, Brockett's steepest descent equation and constrained flows*; in Proceedings of MTNS (C. Byrnes and A. Lindquist, eds.), Springer.

Bloch, A. [1990] *Steepest descent, linear programming and Hamiltonian flows*; in Contemporary Mathematics, **114**, 77–90.

Brockett, R.W. [1991] *Dynamical systems that sort lists, diagonalize matrices and solve linear programming problems*, Linear algebra and its applications, **146**, pp. 79–91 (see also Proceedings of the 1988 IEEE Conference on Decision and Control, pp. 798–803).

Brockett, R.W. *Differential Geometry and the design of gradient algorithms*, (to appear).

Brockett, R.W., and Wong, W.S. [1991] *A gradient flow for the Assignment problem*; in Progress in System and Control Theory (P. Conte and B. Wyman, eds.), Birhauser.

Brøndsted, A. [1983] *Introduction to convex polytopes*, Springer.

Chu, M. [1988] *On the continuous realization of iterative processes*, SIAM Review, **30**, 375-389.

Deift, P., Nanda, T., and Tomei, C. [1983] *Differential Equations for the symmetric eigenvalue problem*, SIAM J. on Numerical Analysis, **20**, 1–22.

Faybusovich, L [1991a] *Dynamical systems which solve optimization problems with linear constraints*, IMA J. of Mathematical Control and Information, **8**, 135–149

———— [1991b] *Hamiltonian structure of dynamical systems which solve linear programming problems*, Physica D, **53**, 217–232.

———— [1991c] *Interior-point methods and Entropy*, Proceedings of the 1991 IEEE Conference on Decision and Control, pp. 2094–2095.

______ [1991d] *Dynamical systems which solve eigenvalue and linear programming problems*, Ph.D. thesis, Harvard University.

______ *Dynamical systems which solve linear programming problems*, Proceedings of the 1992 IEEE Conference on Decision and Control, (to appear).

Gonzaga, C. [1992] *Path-following methods for linear programming*, SIAM Review, **34**, 167–224.

Goldfarb, D., and Todd, M. [1989] *Linear programming*; in Optimization, Vol. 1 (G. Nemhauser, A.R. Kan and M. Todd, eds.), North-Holland.

Lanford, O. [1973] *Entropy and equilibrium states in classical statistical mechanics*; in Lecture notes in physics, **20**, 1–113.

Watkins, D. [1982] *Understanding the QR-algorithm*, SIAM Review, **24**, 427–440.

Submitted for publication February, 1992.

Fields Institute Communications
Volume **3**, 1994

Optimization Techniques on Riemannian Manifolds

Steven T. Smith
Massachusetts Institute of Technology
Lexington, Massachusetts, USA
02173

Abstract. The techniques and analysis presented in this paper provide new methods to solve optimization problems posed on Riemannian manifolds. A new point of view is offered for the solution of constrained optimization problems. Some classical optimization techniques on Euclidean space are generalized to Riemannian manifolds. Several algorithms are presented and their convergence properties are analyzed employing the Riemannian structure of the manifold. Specifically, two apparently new algorithms, which can be thought of as Newton's method and the conjugate gradient method on Riemannian manifolds, are presented and shown to possess, respectively, quadratic and superlinear convergence. Examples of each method on certain Riemannian manifolds are given with the results of numerical experiments. Rayleigh's quotient defined on the sphere is one example. It is shown that Newton's method applied to this function converges cubically, and that the Rayleigh quotient iteration is an efficient approximation of Newton's method. The Riemannian version of the conjugate gradient method applied to this function gives a new algorithm for finding the eigenvectors corresponding to the extreme eigenvalues of a symmetric matrix. Another example arises from extremizing the function $\operatorname{tr}\Theta^{T}Q\Theta N$ on the special orthogonal group. In a similar example, it is shown that Newton's method applied to the sum of the squares of the off-diagonal entries of a symmetric matrix converges cubically.

1991 *Mathematics Subject Classification*. Primary 49K10; Secondary 34A34.

The author enthusiastically thanks Tony Bloch and The Fields Institute for the invitation to speak at The Fields Institute and for their generous support during his visit. The author also thanks Roger Brockett for his suggestion to investigate conjugate gradient methods on manifolds and for his criticism of this work, and the referee for his helpful suggestions. This work was supported in part by the National Science Foundation under the Engineering Research Center Program NSF D CRD-8803012, the Army Research Office under Grant DAA103-92-G-0164 supporting the Brown, Harvard, and MIT Center for Intelligent Control, and by DARPA under Air Force contract F49620-92-J-0466.

Supported in part by the Ministry of Colleges and Universities of Ontario and the Natural Sciences and Engineering Research Council of Canada while visiting The Fields Institute.

1 Introduction

The preponderance of optimization techniques address problems posed on Euclidean spaces. Indeed, several fundamental algorithms have arisen from the desire to compute the minimum of quadratic forms on Euclidean space. However, many optimization problems are posed on non-Euclidean spaces. For example, finding the largest eigenvalue of a symmetric matrix may be posed as the maximization of Rayleigh's quotient defined on the sphere. Optimization problems subject to nonlinear differentiable equality constraints on Euclidean space also lie within this category. Many optimization problems share with these examples the structure of a differentiable manifold endowed with a Riemannian metric. This is the subject of this paper: the extremization of functions defined on Riemannian manifolds.

The minimization of functions on a Riemannian manifold is, at least locally, equivalent to the smoothly constrained optimization problem on a Euclidean space, because every C^∞ Riemannian manifold can be isometrically imbedded in some Euclidean space (Spivak [1979], Vol. 5). However, the dimension of the Euclidean space may be larger than the dimension of the manifold; practical and aesthetic considerations suggest that one try to exploit the intrinsic structure of the manifold. Elements of this spirit may be found throughout the field of numerical methods, such as the emphasis on unitary (norm preserving) transformations in numerical linear algebra (Golub and Van Loan [1983]), or the use of feasible direction methods (Fletcher [1987], Gill and Murray [1974], Sargent [1974]).

An intrinsic approach leads one from the extrinsic idea of vector addition to the exponential map and parallel translation, from minimization along lines to minimization along geodesics, and from partial differentiation to covariant differentiation. The computation of geodesics, parallel translation, and covariant derivatives can be quite expensive. For an n-dimensional manifold, the computation of geodesics and parallel translation requires the solution of a system of $2n$ nonlinear and n linear ordinary differential equations. Nevertheless, many optimization problems are posed on manifolds that have an underlying algebraic structure that may be exploited to greatly reduce the complexity of these computations. For example, on a real compact semisimple Lie group endowed with its natural Riemannian metric, geodesics and parallel translation may be computed via matrix exponentiation (Helgason [1978]). Several algorithms are available to perform this computation (Golub and Van Loan [1983], Moler and Van Loan [1978]). This algebraic structure may be found in the problems posed by Brockett [1989], Brockett [1991], Brockett, (to appear), Bloch, Brockett and Ratiu [1990], Bloch, Brockett and Ratiu [1992], Smith [1991], Faybusovich [1991], Lagarias [1991], Chu [1986], Chu and Driessel [1990], Perkins, Helmke and Moore [1990], and Helmke [1991]. This approach is also applicable if the manifold can be identified with a symmetric space or, excepting parallel translation, a reductive homogeneous space (Kobayashi and Nomizu [1969], Nomizu [1954]). Perhaps the simplest nontrivial example is the sphere, where geodesics and parallel translation can be computed at low cost with trigonometric functions and vector addition. Furthermore, Brown and Bartholomew-Biggs [1989] show that in some cases function minimization by following the solution of a system of ordinary differential equations can be implemented such that it is competitive with conventional techniques.

The outline of the paper is as follows. In Section 2, the optimization problem is posed and conventions to be held throughout the paper are established. The method of steepest descent on a Riemannian manifold is described in Section 3. To fix ideas, a proof of linear convergence is given. The examples of Rayleigh's quotient on the sphere and the function $\operatorname{tr}\Theta^{\mathrm{T}} Q\Theta N$ on the special orthogonal group are presented. In Section 4, Newton's method on a Riemannian manifold is derived. As in Euclidean space, this algorithm may be used to compute the extrema of differentiable functions. It is proved that this method converges quadratically. The example of Rayleigh's quotient is continued, and it is shown that Newton's method applied to this function converges cubically, and is approximated by the Rayleigh quotient iteration. The example considering $\operatorname{tr}\Theta^{\mathrm{T}} Q\Theta N$ is continued. In a related example, it is shown that Newton's method applied to the sum of the squares of the off-diagonal elements of a symmetric matrix converges cubically. This provides an example of a cubically convergent Jacobi-like method. The conjugate gradient method is presented in Section 5 with a proof of superlinear convergence. This technique is shown to provide an effective algorithm for computing the extreme eigenvalues of a symmetric matrix. The conjugate gradient method is applied to the function $\operatorname{tr}\Theta^{\mathrm{T}} Q\Theta N$.

2 Preliminaries

This paper is concerned with the following problem.

Problem 2.1 *Let M be a complete Riemannian manifold, and f a C^∞ function on M. Compute*

$$\min_{p\in M} f(p).$$

There are many well-known algorithms for solving this problem in the case where M is a Euclidean space. This paper generalizes several of these algorithms to the case of complete Riemannian manifolds by replacing the Euclidean notions of straight lines and ordinary differentiation with geodesics and covariant differentiation. These concepts are reviewed in the following paragraphs. We follow Helgason's (Helgason [1978]) and Spivak's (Spivak [1979]) treatments of covariant differentiation, the exponential map, and parallel translation. Details may be found in these references.

Let M be a complete n-dimensional Riemannian manifold with Riemannian structure g and corresponding Levi-Civita connection ∇. Denote the tangent plane at p in M by T_p or $T_p M$. For every p in M, the Riemannian structure g provides an inner product on T_p given by the nondegenerate symmetric bilinear form $g_p\colon T_p \times T_p \to \mathbf{R}$. The notation $\langle X, Y\rangle = g_p(X, Y)$ and $\|X\| = g_p(X, X)^{1/2}$, where X, $Y \in T_p$, is often used. The distance between two points p and q in M is denoted by $d(p, q)$. The gradient of a real-valued C^∞ function f on M at p, denoted by $(\operatorname{grad} f)_p$, is the unique vector in T_p such that $df_p(X) = \langle (\operatorname{grad} f)_p, X\rangle$ for all X in T_p.

Denote the set of C^∞ functions on M by $C^\infty(M)$ and the set of C^∞ vector fields on M by $\mathfrak{X}(M)$. An affine connection on M is a function ∇ which assigns to each vector field $X \in \mathfrak{X}(M)$ an $\mathbf{R}$-linear map $\nabla_X\colon \mathfrak{X}(M) \to \mathfrak{X}(M)$ which satisfies

$$\text{(i)} \quad \nabla_{fX+gY} = f\nabla_X + g\nabla_Y, \qquad \text{(ii)} \quad \nabla_X(fY) = f\nabla_X Y + (Xf)Y,$$

for all f, $g \in C^\infty(M)$, X, $Y \in \mathfrak{X}(M)$. The map ∇_X may be applied to tensors of arbitrary type. Let ∇ be an affine connection on M and $X \in \mathfrak{X}(M)$. Then there exists a unique $\mathbf{R}$-linear map $A \mapsto \nabla_X A$ of C^∞ tensor fields into C^∞ tensor fields which satisfies

(i) $\nabla_X f = Xf$,   (iv) ∇_X preserves the type of tensors,

(ii) $\nabla_X Y$ is given by ∇,   (v) ∇_X commutes with contractions,

(iii) ∇_X is a derivation: $\nabla_X(A \otimes B) = \nabla_X A \otimes B + A \otimes \nabla_X B$,

where $f \in C^\infty(M)$, $Y \in \mathfrak{X}(M)$, and A, B are C^∞ tensor fields. If A is of type (k, l), then $\nabla_X A$, called the covariant derivative of A along X, is of type (k, l), and $\nabla A \colon X \mapsto \nabla_X A$, called the covariant differential of A, is of type $(k, l+1)$.

Let M be a differentiable manifold with affine connection ∇. Let $\gamma \colon I \to M$ be a smooth curve with tangent vectors $X(t) = \dot\gamma(t)$, where $I \subset \mathbf{R}$ is an open interval. The curve γ is called a geodesic if $\nabla_X X = 0$ for all $t \in I$. Let $Y(t) \in T_{\gamma(t)}$ $(t \in I)$ be a smooth family of tangent vectors defined along γ. The family $Y(t)$ is said to be parallel along γ if $\nabla_X Y = 0$ for all $t \in I$.

For every p in M and $X \neq 0$ in T_p, there exists a unique geodesic $t \mapsto \gamma_X(t)$ such that $\gamma_X(0) = p$ and $\dot\gamma_X(0) = X$. We define the exponential map $\exp_p \colon T_p \to M$ by $\exp_p(X) = \gamma_X(1)$ for all $X \in T_p$ such that 1 is in the domain of γ_X. Oftentimes the map $\exp_p$ will be denoted by "exp" when the choice of tangent plane is clear, and $\gamma_X(t)$ will be denoted by $\exp tX$. A neighborhood N_p of p in M is a normal neighborhood if $N_p = \exp N_0$, where N_0 is a star-shaped neighborhood of the origin in T_p and $\exp$ maps N_0 diffeomorphically onto N_p. Normal neighborhoods always exist.

Given a curve $\gamma \colon I \to M$ such that $\gamma(0) = p$, for each $Y \in T_p$ there exists a unique family $Y(t) \in T_{\gamma(t)}$ $(t \in I)$ of tangent vectors parallel along γ such that $Y(0) = Y$. If γ joins the points p and $\gamma(\alpha) = q$, the parallelism along γ induces an isomorphism $\tau_{pq} \colon T_p \to T_q$ defined by $\tau_{pq} Y = Y(\alpha)$.

Let M be a manifold with an affine connection ∇, and N_p a normal neighborhood of $p \in M$. Define the vector field $\tilde{X}$ on N_p adapted to the tangent vector X in T_p by putting $\tilde{X}_q = \tau_{pq} X$, the parallel translation of X along the unique geodesic segment joining p and q.

Given a Riemannian structure g on M, there exists a unique affine connection ∇ on M, called the Levi-Civita connection, which for all X, $Y \in \mathfrak{X}(M)$ satisfies

(i) $\nabla_X Y - \nabla_Y X = [X, Y]$ (∇ is symmetric or torsion-free),

(ii) $\nabla g = 0$ (parallel translation is an isometry).

Length minimizing curves on M are geodesics of the Levi-Civita connection. We shall use this connection throughout the paper.

Unless otherwise specified, all manifolds, vector fields, and functions are assumed to be smooth. When considering a function f to be minimized, the assumption that f is differentiable of class C^∞ can be relaxed throughout the paper, but f must be continuously differentiable at least beyond the derivatives that appear. As the results of this paper are local ones, the assumption that M be complete may also be relaxed in certain instances.

We will use the the following definitions to compare the convergence rates of various algorithms.

Definition 2.2 *Let $\{p_i\}$ be a Cauchy sequence in M that converges to $\hat{p}$. (i) The sequence $\{p_i\}$ is said to converge (at least) linearly if there exists an integer N and a constant $\theta \in [0,1)$ such that $d(p_{i+1}, \hat{p}) \leq \theta d(p_i, \hat{p})$ for all $i \geq N$. (ii) The sequence $\{p_i\}$ is said to converge (at least) quadratically if there exists an integer N and a constant $\theta \geq 0$ such that $d(p_{i+1}, \hat{p}) \leq \theta d^2(p_i, \hat{p})$ for all $i \geq N$. (iii) The sequence $\{p_i\}$ is said to converge (at least) cubically if there exists an integer N and a constant $\theta \geq 0$ such that $d(p_{i+1}, \hat{p}) \leq \theta d^3(p_i, \hat{p})$ for all $i \geq N$. (iv) The sequence $\{p_i\}$ is said to converge superlinearly if it converges faster than any sequence that converges linearly.*

3 Steepest descent on Riemannian manifolds

The method of steepest descent on a Riemannian manifold is conceptually identical to the method of steepest descent on Euclidean space. Each iteration involves a gradient computation and minimization along the geodesic determined by the gradient. Fletcher [1987], Botsaris [1978a], Botsaris [1978b], Botsaris [1981], and Luenberger [1973] describe this algorithm in Euclidean space. Gill and Murray [1974] and Sargent [1974] apply this technique in the presence of constraints. In this section we restate the method of steepest descent described in the literature and provide an alternative formalism that will be useful in the development of Newton's method and the conjugate gradient method on Riemannian manifolds.

Algorithm 3.1 (The method of steepest descent) *Let M be a complete Riemannian manifold with Riemannian structure g and Levi-Civita connection ∇, and let $f \in C^\infty(M)$.*
    Step 0: *Select $p_0 \in M$, compute $G_0 = -(\mathrm{grad}f)_{p_0}$, and set $i = 0$.*

    Step 1: *Compute λ_i such that*

$$f(\exp_{p_i} \lambda_i G_i) \leq f(\exp_{p_i} \lambda G_i)$$

for all $\lambda \geq 0$.

    Step 2: *Set*

$$p_{i+1} = \exp_{p_i} \lambda_i G_i,$$
$$G_{i+1} = -(\mathrm{grad}f)_{p_{i+1}},$$

increment i, and go to Step 1.

It is easy to verify that $\langle G_{i+1}, \tau G_i \rangle = 0$, for $i \geq 0$, where τ is the parallelism with respect to the geodesic from p_i to p_{i+1}. By assumption, the function $\lambda \mapsto f(\exp \lambda G_i)$ is minimized at λ_i. Therefore, we have $0 = (d/dt)|_{t=0} f(\exp(\lambda_i + t)G_i) = df_{p_{i+1}}(\tau G_i) = \langle (\mathrm{grad}f)_{p_{i+1}}, \tau G_i \rangle$. Thus the method of steepest descent on a Riemannian manifold has the same deficiency as its counterpart on a Euclidean space, i.e., it makes a ninety degree turn at every step.

The convergence of Algorithm 3.1 is linear. To prove this fact, we will make use of a standard theorem of the calculus, expressed in differential geometric language. The covariant derivative $\nabla_X f$ of f along X is defined to be Xf. For $k = 1, 2, \ldots$, define $\nabla_X^k f = \nabla_X \circ \cdots \circ \nabla_X f$ (k times), and let $\nabla_X^0 f = f$.

Remark 3.2 (Taylor's formula) Let M be a manifold with an affine connection ∇, N_p a normal neighborhood of $p \in M$, the vector field $\tilde{X}$ on N_p adapted to X in T_p, and f a C^∞ function on M. Then there exists an $\epsilon > 0$ such that for every $\lambda \in [0, \epsilon)$

$$
f(\exp_p \lambda X) = f(p) + \lambda(\nabla_{\tilde{X}} f)(p) + \cdots + \frac{\lambda^{n-1}}{(n-1)!}(\nabla_{\tilde{X}}^{n-1} f)(p)
$$
$$
+ \frac{\lambda^n}{(n-1)!} \int_0^1 (1-t)^{n-1} (\nabla_{\tilde{X}}^n f)(\exp_p t\lambda X)\, dt. \tag{3.1}
$$

Proof Let N_0 be a star-shaped neighborhood of $0 \in T_p$ such that $N_p = \exp N_0$. There exists $\epsilon > 0$ such that $\lambda X \in N_0$ for all $\lambda \in [0, \epsilon)$. The map $\lambda \mapsto f(\exp \lambda X)$ is a real C^∞ function on $[0, \epsilon)$ with derivative $(\nabla_{\tilde{X}} f)(\exp \lambda X)$. The statement follows by repeated integration by parts. $\square$

The following special cases of Remark 3.2 will be particularly useful. When $n = 2$, Equation (3.1) yields

$$
f(\exp_p \lambda X) = f(p) + \lambda(\nabla_{\tilde{X}} f)(p) + \lambda^2 \int_0^1 (1-t)(\nabla_{\tilde{X}}^2 f)(\exp_p t\lambda X)\, dt. \tag{3.2}
$$

Furthermore, when $n = 1$, Equation (3.1) applied to the function $\tilde{X} f = \nabla_{\tilde{X}} f$ yields

$$
(\tilde{X} f)(\exp_p \lambda X) = (\tilde{X} f)(p) + \lambda \int_0^1 (\nabla_{\tilde{X}}^2 f)(\exp_p t\lambda X)\, dt. \tag{3.3}
$$

The convergence proofs require a characterization of the second order terms of f near a critical point. Consider the second covariant differential $\nabla\nabla f = \nabla^2 f$ of a smooth function $f: M \to \mathbf{R}$. If $(U, x^1, \dots, x^n)$ is a coordinate chart on M, then at $p \in U$ this $(0,2)$ tensor takes the form

$$
(\nabla^2 f)_p = \sum_{i,j} \left(\left(\frac{\partial^2 f}{\partial x^i \partial x^j} \right)_p - \sum_k \Gamma_{ji}^k \left(\frac{\partial f}{\partial x^k} \right)_p \right) dx^i \otimes dx^j \tag{3.4}
$$

where Γ_{ij}^k are the Christoffel symbols at p. If $\hat{p}$ in U is a critical point of f, then $(\partial f / \partial x^k)_{\hat{p}} = 0$, $k = 1, \dots, n$. Therefore $(\nabla^2 f)_{\hat{p}} = (d^2 f)_{\hat{p}}$, where $(d^2 f)_{\hat{p}}$ is the Hessian of f at the critical point $\hat{p}$. Furthermore, for $p \in M$, $X, Y \in T_p$, and $\tilde{X}$ and $\tilde{Y}$ vector fields adapted to X and Y, respectively, on a normal neighborhood N_p of p, we have $(\nabla^2 f)(\tilde{X}, \tilde{Y}) = \nabla_{\tilde{Y}} \nabla_{\tilde{X}} f$ on N_p. Therefore the coefficient of the second term of the Taylor expansion of $f(\exp tX)$ is $(\nabla_{\tilde{X}}^2 f)_p = (\nabla^2 f)_p(X, X)$. Note that the bilinear form $(\nabla^2 f)_p$ on $T_p \times T_p$ is symmetric if and only if ∇ is symmetric, which true of the Levi-Civita connection by definition.

Theorem 3.3 *Let M be a complete Riemannian manifold with Riemannian structure g and Levi-Civita connection ∇. Let $f \in C^\infty(M)$ have a nondegenerate critical point at $\hat{p}$ such that the Hessian $(d^2 f)_{\hat{p}}$ is positive definite. Let p_i be a sequence of points in M converging to $\hat{p}$ and $H_i \in T_{p_i}$ a sequence of tangent vectors such that*

$$
\text{(i)} \qquad p_{i+1} = \exp_{p_i} \lambda_i H_i \qquad\qquad \text{for } i = 0, 1, \dots,
$$
$$
\text{(ii)} \quad \langle -(\operatorname{grad} f)_{p_i}, H_i \rangle \geq c \|(\operatorname{grad} f)_{p_i}\| \, \|H_i\| \quad \text{for } c \in (0, 1],
$$

where λ_i is chosen such that $f(\exp \lambda_i H_i) \leq f(\exp \lambda H_i)$ for all $\lambda \geq 0$. Then there exists a constant E and a $\theta \in [0, 1)$ such that for all $i = 0, 1, \ldots ,$

$$d(p_i, \hat{p}) \leq E\theta^i.$$

Proof The proof is a generalization of the one given in Polak [1971], p. 242ff for the method of steepest descent on Euclidean space.

The existence of a convergent sequence is guaranteed by the smoothness of f. If $p_j = \hat{p}$ for some integer j, the assertion becomes trivial; assume otherwise. By the smoothness of f, there exists an open neighborhood U of $\hat{p}$ such that $(\nabla^2 f)_p$ is positive definite for all $p \in U$. Therefore, there exist constants $k > 0$ and $K \geq k > 0$ such that for all $X \in T_p$ and all $p \in U$,

$$k\|X\|^2 \leq (\nabla^2 f)_p(X, X) \leq K\|X\|^2. \tag{3.5}$$

Define $X_i \in T_{\hat{p}}$ by the relations $\exp X_i = p_i$, $i = 0, 1, \ldots$. By assumption, $df_{\hat{p}} = 0$ and from Equation (3.2), we have

$$f(p_i) - f(\hat{p}) = \int_0^1 (1 - t)(\nabla^2_{\tilde{X}_i} f)(\exp_{\hat{p}} tX_i)\, dt. \tag{3.6}$$

Combining this equality with the inequalities of (3.5) yields

$$\tfrac{1}{2} k d^2(p_i, \hat{p}) \leq f(p_i) - f(\hat{p}) \leq \tfrac{1}{2} K d^2(p_i, \hat{p}). \tag{3.7}$$

Similarly, we have by Equation (3.3)

$$(\tilde{X}_i f)(p_i) = \int_0^1 (\nabla^2_{\tilde{X}_i} f)(\exp_{\hat{p}} tX_i)\, dt.$$

Next, use (3.6) with Schwarz's inequality and the first inequality of (3.7) to obtain

$$kd^2(p_i, \hat{p}) = k\|X_i\|^2 \leq \int_0^1 (\nabla^2_{\tilde{X}_i} f)(\exp_{\hat{p}} tX_i)\, dt = (\tilde{X}_i f)(p_i)$$
$$= df_{p_i}\big((\tilde{X}_i)_{p_i}\big) = df_{p_i}(\tau X_i) = \langle (\mathrm{grad} f)_{p_i}, \tau X_i \rangle$$
$$\leq \|(\mathrm{grad} f)_{p_i}\|\, \|\tau X_i\| = \|(\mathrm{grad} f)_{p_i}\|\, d(p_i, \hat{p}).$$

Therefore,

$$\|(\mathrm{grad} f)_{p_i}\| \geq k d(p_i, \hat{p}). \tag{3.8}$$

Define the function $\Delta \colon T_p \times \mathbf{R} \to \mathbf{R}$ by the equation $\Delta(X, \lambda) = f(\exp_p \lambda X) - f(p)$. By Equation (3.2), the second order Taylor formula, we have

$$\Delta(H_i, \lambda) = \lambda(\tilde{H}_i f)(p_i) + \tfrac{1}{2}\lambda^2 \int_0^1 (1 - t)(\nabla^2_{\tilde{H}_i} f)(\exp_{p_i} \lambda H_i)\, dt.$$

Using assumption (ii) of the theorem along with (3.5) we establish for $\lambda \geq 0$

$$\Delta(H_i, \lambda) \leq -\lambda c\|(\mathrm{grad} f)_{p_i}\|\, \|H_i\| + \tfrac{1}{2}\lambda^2 K\|H_i\|^2. \tag{3.9}$$

We may now compute an upper bound for the rate of linear convergence θ. By assumption (i) of the theorem, λ must be chosen to minimize the right hand side of (3.9). This corresponds to choosing $\lambda = c\|(\mathrm{grad} f)_{p_i}\| / K\|H_i\|$. A computation reveals that

$$\Delta(H_i, \lambda_i) \leq -\frac{c^2}{2K}\|(\mathrm{grad} f)_{p_i}\|^2.$$

Applying (3.7) and (3.8) to this inequality and rearranging terms yields

$$f(p_{i+1}) - f(\hat{p}) \leq \theta\big(f(p_i) - f(\hat{p})\big), \tag{3.10}$$

where $\theta = \big(1 - (ck/K)^2\big)$. By assumption, $c \in (0,1]$ and $0 < k \leq K$, therefore $\theta \in [0,1)$. (Note that Schwarz's inequality bounds c below unity.) From (3.10) it is seen that $\big(f(p_i) - f(\hat{p})\big) \leq E\theta^i$ where $E = \big(f(p_0) - f(\hat{p})\big)$. From (3.7) we conclude that for $i = 0, 1, \ldots,$

$$d(p_i, \hat{p}) \leq \sqrt{\frac{2E}{k}}\,\big(\sqrt{\theta}\,\big)^i. \tag{3.11}$$

$\square$

Corollary 3.4 *If Algorithm 3.1 converges to a local minimum, it converges linearly.*

The choice $H_i = -(\mathrm{grad}\,f)_{p_i}$ yields $c = 1$ in the second assumption the Theorem 3.3, which establishes the corollary.

Example 3.5 (Rayleigh's quotient on the sphere) *Let S^{n-1} be the imbedded sphere in $\mathbf{R}^n$, i.e., $S^{n-1} = \{\,x \in \mathbf{R}^n : x^{\mathrm{T}}x = 1\,\}$, where $x^{\mathrm{T}}y$ denotes the standard inner product on $\mathbf{R}^n$, which induces a metric on S^{n-1}. Geodesics on the sphere are great circles and parallel translation along geodesics is equivalent to rotating the tangent plane along the great circle. Let $x \in S^{n-1}$ and $h \in T_x$ have unit length, and $v \in T_x$ be any tangent vector. Then*

$$\exp_x th = x \cos t + h \sin t,$$
$$\tau h = h \cos t - x \sin t,$$
$$\tau v = v - (h^{\mathrm{T}}v)\big(x \sin t + h(1 - \cos t)\big),$$

where τ is the parallelism along the geodesic $t \mapsto \exp th$. Let Q be an n-by-n positive definite symmetric matrix with distinct eigenvalues and define $\rho \colon S^{n-1} \to \mathbf{R}$ by $\rho(x) = x^{\mathrm{T}}Qx$. A computation shows that

$$\tfrac{1}{2}(\mathrm{grad}\,\rho)_x = Qx - \rho(x)x. \tag{3.12}$$

The function ρ has a unique minimum and maximum point at the eigenvectors corresponding to the smallest and largest eigenvalues of Q, respectively. Because S^{n-1} is geodesically complete, the method of steepest descent in the opposite direction of the gradient converges to the eigenvector corresponding to the smallest eigenvalue of Q; likewise for the eigenvector corresponding to the largest eigenvalue. Chu [1986] considers the continuous limit of this problem. A computation shows that $\rho(x)$ is maximized along the geodesic $\exp_x th$ ($\|h\| = 1$) when $a \cos 2t - b \sin 2t = 0$, where $a = 2x^{\mathrm{T}}Qh$ and $b = \rho(x) - \rho(h)$. Thus $\cos t$ and $\sin t$ may be computed with simple algebraic functions of a and b (which appear below in Algorithm 5.5). The results of a numerical experiment demonstrating the convergence of the method of steepest descent applied to maximizing Rayleigh's quotient on S^{20} are shown in Figure 1.

Example 3.6 (Brockett [1991], Brockett, (to appear)) *Consider the function $f(\Theta) = \mathrm{tr}\,\Theta^{\mathrm{T}}Q\Theta N$ on the special orthogonal group $SO(n)$, where Q is a real symmetric matrix with distinct eigenvalues and N is a real diagonal matrix with distinct diagonal elements. It will be convenient to identify tangent vectors in T_Θ with* *tangent vectors in $T_I \cong \mathfrak{so}(n)$, the tangent plane at the identity, via left translation. The gradient of f (with respect to the negative Killing form of $\mathfrak{so}(n)$, scaled by $1/(n-2)$) at $\Theta \in SO(n)$ is $\Theta[H, N]$, where $H = \mathrm{Ad}_{\Theta^T}(Q) = \Theta^T Q \Theta$. The group $SO(n)$ acts on the set of symmetric matrices by conjugation; the orbit of Q under the action of $SO(n)$ is an isospectral submanifold of the symmetric matrices. We seek a $\hat{\Theta}$ such that $f(\hat{\Theta})$ is maximized. This point corresponds to a diagonal matrix whose diagonal entries are ordered similarly to those of N. A related example is found in Smith [1991], who considers the homogeneous space of matrices with fixed singular values, and in Chu and Driessel [1990].*

The Levi-Civita connection on $SO(n)$ is bi-invariant and invariant with respect to inversion; therefore, geodesics and parallel translation may be computed via matrix exponentiation of elements in $\mathfrak{so}(n)$ and left (or right) translation (Helgason [1978], Ch. 2, Ex. 6). The geodesic emanating from the identity in $SO(n)$ in direction $X \in \mathfrak{so}(n)$ is given by the formula $\exp_I tX = e^{tX}$, where the right hand side denotes regular matrix exponentiation. The expense of geodesic minimization may be avoided if instead one uses Brockett's estimate (Brockett, (to appear)) for the step size. Given $\Omega \in \mathfrak{so}(n)$, we wish to find $t > 0$ such that $\phi(t) = \mathrm{tr}\, \mathrm{Ad}_{e^{-t\Omega}}(H)N$ is minimized. Differentiating ϕ twice shows that $\phi'(t) = -\mathrm{tr}\, \mathrm{Ad}_{e^{-t\Omega}}(\mathrm{ad}_\Omega H)N$ and $\phi''(t) = -\mathrm{tr}\, \mathrm{Ad}_{e^{-t\Omega}}(\mathrm{ad}_\Omega H)\,\mathrm{ad}_\Omega N$, where $\mathrm{ad}_\Omega A = [\Omega, A]$. Hence, $\phi'(0) = 2\,\mathrm{tr}\, H\Omega N$ and, by Schwarz's inequality and the fact that Ad is an isometry, $|\phi''(t)| \leq \|\mathrm{ad}_\Omega H\|\, \|\mathrm{ad}_\Omega N\|$. We conclude that if $\phi'(0) > 0$, then ϕ' is nonnegative on the interval

$$0 \leq t \leq \frac{2\,\mathrm{tr}\, H\Omega N}{\|\mathrm{ad}_\Omega H\|\, \|\mathrm{ad}_\Omega N\|}, \tag{3.13}$$

which provides an estimate for the step size of Step 1 in Algorithm 3.1. The results of a numerical experiment demonstrating the convergence of the method of steepest descent (ascent) in $SO(20)$ using this estimate are shown in Figure 2.

4 Newton's method on Riemannian manifolds

As in the optimization of functions on Euclidean space, quadratic convergence can be obtained if the second order terms of the Taylor expansion are used appropriately. In this section we present Newton's algorithm on Riemannian manifolds, prove that its convergence is quadratic, and provide examples. Whereas the convergence proof for the method of steepest descent relies upon the Taylor expansion of the function f, the convergence proof for Newton's method will rely upon the Taylor expansion of the one-form df. Note that Newton's method has a counterpart in the theory of constrained optimization, as described by, e.g., Fletcher [1987], Bertsekas [1982a], Bertsekas [1982b], Dunn [1980], or Dunn [1981]. The Newton method presented in this section has only local convergence properties. There is a theory of global Newton methods on Euclidean space and computational complexity; see the work of Hirsch and Smale [1979], Smale [1981], Smale [1985], Shub and Smale [1985], and Shub and Smale [1986a].

Let M be an n-dimensional Riemannian manifold with Riemannian structure g and Levi-Civita connection ∇, let μ be a C^∞ one-form on M, and let p in M be such that the bilinear form $(\nabla\mu)_p\colon T_p \times T_p \to \mathbf{R}$ is nondegenerate. Then, by abuse of notation, we have the pair of isomorphisms

$$T_p \xrightleftharpoons[(\nabla\mu)_p^{-1}]{(\nabla\mu)_p} T_p^*$$

with the forward map defined by $X \mapsto (\nabla_X \mu)_p = (\nabla\mu)_p(\cdot, X)$, which is nonsingular. The notation $(\nabla\mu)_p$ will henceforth be used for both the bilinear form defined by the covariant differential of μ evaluated at p and the homomorphism from T_p to T_p^* induced by this bilinear form. In case of an isomorphism, the inverse can be used to compute a point in M where μ vanishes, if such a point exists. The case $\mu = df$ will be of particular interest, in which case $\nabla\mu = \nabla^2 f$. Before expounding on these ideas, we make the following remarks.

Remark 4.1 (The mean value theorem) Let M be a manifold with affine connection ∇, N_p a normal neighborhood of $p \in M$, the vector field $\tilde{X}$ on N_p adapted to $X \in T_p$, μ a one-form on N_p, and τ_λ the parallelism with respect to $\exp tX$ for $t \in [0, \lambda]$. Denote the point $\exp \lambda X$ by p_λ. Then there exists an $\epsilon > 0$ such that for every $\lambda \in [0, \epsilon)$, there is an $\alpha \in [0, \lambda]$ such that

$$\tau_\lambda^{-1} \mu_{p_\lambda} - \mu_p = \lambda (\nabla_{\tilde{X}} \mu)_{p_\alpha} \circ \tau_\alpha.$$

Proof As in the proof of Remark 3.2, there exists an $\epsilon > 0$ such that $\lambda X \in N_0$ for all $\lambda \in [0, \epsilon)$. The map $\lambda \mapsto (\tau_\lambda^{-1} \mu_{p_\lambda})(A)$, for any A in T_p, is a C^∞ function on $[0, \epsilon)$ with derivative $(d/dt)(\tau_t^{-1} \mu_{p_t})(A) = (d/dt)\mu_{p_t}(\tau_t A) = \nabla_{\tilde{X}}\big(\mu_{p_t}(\tau_t A)\big) = (\nabla_{\tilde{X}}\mu)_{p_t}(\tau_t A) + \mu_{p_t}\big(\nabla_{\tilde{X}}(\tau_t A)\big) = (\nabla_{\tilde{X}}\mu)_{p_t}(\tau_t A)$. The lemma follows from the mean value theorem of real analysis. $\square$

This remark can be generalized in the following way.

Remark 4.2 (Taylor's theorem) Let M be a manifold with affine connection ∇, N_p a normal neighborhood of $p \in M$, the vector field $\tilde{X}$ on N_p adapted to $X \in T_p$, μ a one-form on N_p, and τ_λ the parallelism with respect to $\exp tX$ for $t \in [0, \lambda]$. Denote the point $\exp \lambda X$ by p_λ. Then there exists an $\epsilon > 0$ such that for every $\lambda \in [0, \epsilon)$, there is an $\alpha \in [0, \lambda]$ such that

$$\tau_\lambda^{-1} \mu_{p_\lambda} = \mu_p + \lambda(\nabla_{\tilde{X}}\mu)_p + \cdots + \frac{\lambda^{n-1}}{(n-1)!}(\nabla_{\tilde{X}}^{n-1}\mu)_p + \frac{\lambda^n}{n!}(\nabla_{\tilde{X}}^n\mu)_{p_\alpha} \circ \tau_\alpha. \tag{4.1}$$

The remark follows by applying Remark 4.1 and the Taylor's theorem of real analysis to the function $\lambda \mapsto (\tau_\lambda^{-1} \mu_{p_\lambda})(A)$ for any A in T_p.

Remarks 4.1 and 4.2 can be generalized to C^∞ tensor fields, but we will only require Remark 4.2 for case $n = 2$ to make the following observation.

Let μ be a one-form on M such that for some $\hat{p}$ in M, $\mu_{\hat{p}} = 0$. Given any p in a normal neighborhood of $\hat{p}$, we wish to find X in T_p such that $\exp_p X = \hat{p}$. Consider the Taylor expansion of μ about p, and let τ be the parallel translation along the unique geodesic joining p to $\hat{p}$. We have by our assumption that μ vanishes at $\hat{p}$, and from Equation (4.1) for $n = 2$,

$$0 = \tau^{-1}\mu_{\hat{p}} = \tau^{-1}\mu_{\exp_p X} = \mu_p + (\nabla\mu)_p(\cdot, X) + \text{h.o.t.}$$

If the bilinear form $(\nabla\mu)_p$ is nondegenerate, the tangent vector X may be approximated by discarding the higher order terms and solving the resulting linear equation

$$\mu_p + (\nabla\mu)_p(\cdot, X) = 0$$

for X, which yields

$$X = -(\nabla\mu)_p^{-1}\mu_p.$$

This approximation is the basis of the following algorithm.

Algorithm 4.3 (Newton's method) *Let M be a complete Riemannian manifold with Riemannian structure g and Levi-Civita connection ∇, and let μ be a C^∞ one-form on M.*

Step 0: *Select $p_0 \in M$ such that $(\nabla\mu)_{p_0}$ is nondegenerate, and set $i = 0$.*

Step 1: *Compute*

$$H_i = -(\nabla\mu)_{p_i}^{-1}\mu_{p_i}$$

$$p_{i+1} = \exp_{p_i} H_i,$$

(assume that $(\nabla\mu)_{p_i}$ is nondegenerate), increment i, and repeat.

It can be shown that if p_0 is chosen suitably close (within the so-called domain of attraction) to a point $\hat{p}$ in M such that $\mu_{\hat{p}} = 0$ and $(\nabla\mu)_{\hat{p}}$ is nondegenerate, then Algorithm 4.3 converges quadratically to $\hat{p}$. The following theorem holds for general one-forms; we will consider the case where μ is exact.

Theorem 4.4 *Let $f \in C^\infty(M)$ have a nondegenerate critical point at $\hat{p}$. Then there exists a neighborhood U of $\hat{p}$ such that for any $p_0 \in U$, the iterates of Algorithm 4.3 for $\mu = df$ are well defined and converge quadratically to $\hat{p}$.*

The proof of this theorem is a generalization of the corresponding proof for Euclidean spaces, with an extra term containing the Riemannian curvature tensor (which of course vanishes in the latter case).

Proof If $p_j = \hat{p}$ for some integer j, the assertion becomes trivial; assume otherwise. Define $X_i \in T_{p_i}$ by the relations $\hat{p} = \exp X_i$, $i = 0, 1, \ldots$, so that $d(p_i, \hat{p}) = \|X_i\|$ (n.b. this convention is opposite that used in the proof of Theorem 3.3). Consider the geodesic triangle with vertices p_i, p_{i+1}, and $\hat{p}$, and sides $\exp tX_i$ from p_i to $\hat{p}$, $\exp tH_i$ from p_i to p_{i+1}, and $\exp tX_{i+1}$ from p_{i+1} to $\hat{p}$, for $t \in [0, 1]$. Let τ be the parallelism with respect to the side $\exp tH_i$ between p_i and p_{i+1}. There exists a unique tangent vector Ξ_i in T_{p_i} defined by the equation

$$X_i = H_i + \tau^{-1}X_{i+1} + \Xi_i \tag{4.2}$$

(Ξ_i may be interpreted as the amount by which vector addition fails). If we use the definition $H_i = -(\nabla^2 f)_{p_i}^{-1} df_{p_i}$ of Algorithm 4.3, apply the isomorphism $(\nabla^2 f)_{p_i} : T_{p_i} \to T_{p_i}^*$ to both sides of Equation (4.2), we obtain the equation

$$(\nabla^2 f)_{p_i}(\tau^{-1}X_{i+1}) = df_{p_i} + (\nabla^2 f)_{p_i}X_i - (\nabla^2 f)_{p_i}\Xi_i. \tag{4.3}$$

By Taylor's theorem, there exists an $\alpha \in [0, 1]$ such that

$$\tau_1^{-1}df_{\hat{p}} = df_{p_i} + (\nabla_{\tilde{X}_i}df)_{p_i} + \tfrac{1}{2}(\nabla_{\tilde{X}_i}^2 df)_{p_\alpha} \circ \tau_\alpha \tag{4.4}$$

where τ_t is the parallel translation from p_i to $p_t = \exp tX_i$. The trivial identities $(\nabla_{\tilde{X}_i}df)_{p_i} = (\nabla^2 f)_{p_i}X_i$ and $(\nabla_{\tilde{X}_i}^2 df)_{p_\alpha} = (\nabla^3 f)_{p_\alpha}(\tau_\alpha\cdot, \tau_\alpha X_i, \tau_\alpha X_i)$ will be used to replace the last two terms on the right hand side of Equation (4.4). Combining the assumption that $df_{\hat{p}} = 0$ with Equations (4.3) and (4.4), we obtain

$$(\nabla^2 f)_{p_i}(\tau^{-1} X_{i+1}) = -\tfrac{1}{2}(\nabla^2_{\tilde{X}_i} df)_{p_\alpha} \circ \tau_\alpha - (\nabla^2 f)_{p_i} \Xi_i. \tag{4.5}$$

By the smoothness of f and g, there exists an $\epsilon > 0$ and constants δ', δ'', δ''', all greater than zero, such that whenever p is in the convex normal ball $B_\epsilon(\hat{p})$,

$$
\begin{array}{rll}
\text{(i)} & \|(\nabla^2 f)_p(\cdot, X)\| \geq \delta' \|X\| & \text{for all } X \in T_p, \\[4pt]
\text{(ii)} & \|(\nabla^2 f)_p(\cdot, X)\| \leq \delta'' \|X\| & \text{for all } X \in T_p, \\[4pt]
\text{(iii)} & \|(\nabla^3 f)_p(\cdot, X, X)\| \leq \delta''' \|X\|^2 & \text{for all } X \in T_p,
\end{array}
$$

where the induced norm on T_p^* is used in all three cases. Taking the norm of both sides of Equation (4.5), applying the triangle inequality to the right hand side, and using the fact that parallel translation is an isometry, we obtain the inequality

$$\delta' d(p_{i+1}, \hat{p}) \leq \delta''' d^2(p_i, \hat{p}) + \delta'' \|\Xi_i\|. \tag{4.6}$$

The length of Ξ_i can be bounded by a cubic expression in $d(p_i, \hat{p})$ by considering the distance between the points $\exp(H_i + \tau^{-1} X_{i+1})$ and $\exp X_{i+1} = \hat{p}$. Given $p \in M$, $\epsilon > 0$ small enough, let a, $v \in T_p$ be such that $\|a\| + \|v\| \leq \epsilon$, and let τ be the parallel translation with respect to the geodesic from p to $q = \exp_p a$. Karcher [1977], App. C2.2 shows that

$$d\bigl(\exp_p(a + v), \exp_q(\tau v)\bigr) \leq \|a\| \cdot \text{const.} \, (\max |K|) \cdot \epsilon^2, \tag{4.7}$$

where K is the sectional curvature of M along any section in the tangent plane at any point near p.

There exists a constant $c > 0$ such that $\|\Xi_i\| \leq c \, d\bigl(\hat{p}, \exp(H_i + \tau^{-1} X_{i+1})\bigr)$. By (4.7), we have $\|\Xi_i\| \leq \text{const.} \|H_i\| \epsilon^2$. Taking the norm of both sides of the Taylor formula $df_{p_i} = -\int_0^1 (\nabla_{\tilde{X}_i} df)(\exp t X_i) \, dt$ and applying a standard integral inequality and inequality (ii) from above yields $\|df_{p_i}\| \leq \delta'' \|X_i\|$ so that $\|H_i\| \leq \text{const.} \|X_i\|$. Furthermore, we have the triangle inequality $\|X_{i+1}\| \leq \|X_i\| + \|H_i\|$, therefore ϵ may be chosen such that $\|H_i\| + \|X_{i+1}\| \leq \epsilon \leq \text{const.} \|X_i\|$. By (4.7) there exists $\delta^{\text{iv}} > 0$ such that $\|\Xi_i\| \leq \delta^{\text{iv}} d^3(p_i, \hat{p})$. $\square$

Corollary 4.5 *If $(\nabla^2 f)_{\hat{p}}$ is positive (negative) definite and Algorithm 4.3 converges to $\hat{p}$, then Algorithm 4.3 converges quadratically to a local minimum (maximum) of f.*

Example 4.6 (Rayleigh's quotient on the sphere) *Let S^{n-1} and $\rho(x) = x^{\mathsf{T}} Q x$ be as in Example 3.5. It will be convenient to work with the coordinates $x^1, \ldots, x^n$ of the ambient space $\mathbf{R}^n$, treat the tangent plane $T_x S^{n-1}$ as a vector subspace of $\mathbf{R}^n$, and make the identification $T_x S^{n-1} \cong T_x^* S^{n-1}$ via the metric. In this coordinate system, geodesics on the sphere obey the second order differential equation $\ddot{x}^k + x^k = 0$, $k = 1, \ldots, n$. Thus the Christoffel symbols are given by $\Gamma_{ij}^k = \delta_{ij} x^k$, where δ_{ij} is the Kronecker delta. The ijth component of the second covariant differential of ρ at x in S^{n-1} is given by (cf. Equation (3.4))*

$$\bigl((\nabla^2 \rho)_x\bigr)_{ij} = 2Q_{ij} - \sum_{k,l} \delta_{ij} x^k \cdot 2Q_{kl} x^l = 2\bigl(Q_{ij} - \rho(x)\delta_{ij}\bigr),$$

or, written as matrices,

$$\tfrac{1}{2}(\nabla^2\rho)_x = Q - \rho(x)I. \tag{4.8}$$

Let u be a tangent vector in T_xS^{n-1}. A linear operator $A\colon \mathbf{R}^n \to \mathbf{R}^n$ defines a linear operator on the tangent plane T_xS^{n-1} for each x in S^{n-1} such that

$$A{\cdot}u = Au - (x^{\mathrm{T}}Au)x = (I - xx^{\mathrm{T}})Au.$$

If A is invertible as an endomorphism of the ambient space $\mathbf{R}^n$, the solution to the linear equation $A{\cdot}u = v$ for u, v in T_xS^{n-1} is

$$u = A^{-1}\left(v - \frac{(x^{\mathrm{T}}A^{-1}v)}{(x^{\mathrm{T}}A^{-1}x)}x\right). \tag{4.9}$$

For Newton's method, the direction H_i in T_xS^{n-1} is the solution of the equation

$$(\nabla^2\rho)_{x_i}{\cdot}H_i = -(\operatorname{grad}\rho)_{x_i}.$$

Combining Equations (3.12), (4.8), and (4.9), we obtain

$$H_i = -x_i + \alpha_i\bigl(Q - \rho(x_i)I\bigr)^{-1}x_i$$

where $\alpha_i = 1/x_i^{\mathrm{T}}(Q - \rho(x_i)I)^{-1}x_i$. This gives rise to the following algorithm for computing eigenvectors of the symmetric matrix Q.

Algorithm 4.7 (Newton-Rayleigh quotient method) *Let Q be a real symmetric n-by-n matrix.*

Step 0: *Select x_0 in $\mathbf{R}^n$ such that $x_0^{\mathrm{T}}x_0 = 1$, and set $i = 0$.*

Step 1: *Compute*

$$y_i = \bigl(Q - \rho(x_i)I\bigr)^{-1}x_i$$

and set $\alpha_i = 1/x_i^{\mathrm{T}}y_i$.

Step 2: *Compute*

$$H_i = -x_i + \alpha_iy_i, \quad \theta_i = \|H_i\|,$$
$$x_{i+1} = x_i\cos\theta_i + H_i\sin\theta_i/\theta_i,$$

increment i, and go to Step 1.

The quadratic convergence guaranteed by Theorem 4.4 is in fact too conservative for Algorithm 4.7. As evidenced by Figure 1, Algorithm 4.7 converges cubically.

Proposition 4.8 *If λ is a distinct eigenvalue of the symmetric matrix Q, and Algorithm 4.7 converges to the corresponding eigenvector $\hat{x}$, then it converges cubically.*

Proof [1] In the coordinates $x^1, \dots, x^n$ of the ambient space $\mathbf{R}^n$, the ijkth component of the third covariant differential of ρ at $\hat{x}$ is $-2\lambda\hat{x}^k\delta_{ij}$. Let $X \in T_{\hat{x}}S^{n-1}$. Then $(\nabla^3\rho)_{\hat{x}}(\cdot, X, X) = 0$ and the second order terms on the right hand side of Equation (4.5) vanish at the critical point. The proposition follows from the smoothness of ρ. $\square$

Proof [2] The proof follows Parlett's proof of cubic convergence for the Rayleigh quotient iteration (Parlett [1980], p.72ff). Assume that for all i, $x_i \neq \hat{x}$, and denote $\rho(x_i)$ by ρ_i. For all i, there is an angle ψ_i and a unit length vector u_i defined by the equation $x_i = \hat{x}\cos\psi_i + u_i\sin\psi_i$, such that $\hat{x}^{\mathrm{T}}u_i = 0$. By Algorithm 4.7

$$x_{i+1} = \hat{x}\cos\psi_{i+1} + u_{i+1}\sin\psi_{i+1} = x_i\cos\theta_i + H_i\sin\theta_i/\theta_i$$

$$= \hat{x}\left(\frac{\alpha_i\sin\theta_i}{(\lambda-\rho_i)\theta_i} + \beta_i\right)\cos\psi_i + \left(\frac{\alpha_i\sin\theta_i}{\theta_i}(Q-\rho_i I)^{-1}u_i + \beta_i u_i\right)\sin\psi_i,$$

where $\beta_i = \cos\theta_i - \sin\theta_i/\theta_i$. Therefore,

$$|\tan\psi_{i+1}| = \frac{\left\|\frac{\alpha_i\sin\theta_i}{\theta_i}(Q-\rho_i I)^{-1}u_i + \beta_i u_i\right\|}{\left|\frac{\alpha_i\sin\theta_i}{(\lambda-\rho_i)\theta_i} + \beta_i\right|}\cdot|\tan\psi_i|. \tag{4.10}$$

The following equalities and low order approximations in terms of the small quantities $\lambda - \rho_i$, θ_i, and ψ_i are straightforward to establish: $\lambda - \rho_i = (\lambda - \rho(u_i))\sin^2\psi_i$, $\theta_i^2 = \cos^2\psi_i\sin^2\psi_i + \text{h.o.t.}$, $\alpha_i = (\lambda - \rho_i) + \text{h.o.t.}$, and $\beta_i = -\theta_i^2/3 + \text{h.o.t.}$ Thus, the denominator of the large fraction in Equation (4.10) is of order unity and the numerator is of order $\sin^2\psi_i$. Therefore, we have

$$|\psi_{i+1}| = \text{const.}\,|\psi_i|^3 + \text{h.o.t.} \qquad \square$$

Remark 4.9 If Algorithm 4.7 is simplified by replacing Step 2 with:

Step 2′: *Compute*

$$x_{i+1} = y_i/\|y_i\|,$$

increment i, and go to Step 1.

Then we obtain the Rayleigh quotient iteration. These two algorithms differ by the method in which they use the vector $y_i = (Q - \rho(x_i)I)^{-1}x_i$ to compute the next iterate on the sphere. Algorithm 4.7 computes the point H_i in $T_{x_i}S^{n-1}$ where y_i intersects this tangent plane, then computes x_{i+1} via the exponential map of this vector (which "rolls" the tangent vector H_i onto the sphere). The Rayleigh quotient iteration computes the intersection of y_i with the sphere itself and takes this intersection to be x_{i+1}. The latter approach approximates Algorithm 4.7 up to quadratic terms when x_i is close to an eigenvector. Algorithm 4.7 is more expensive to compute than—though of the same order as—the Rayleigh quotient iteration; thus, the RQI is seen to be an efficient approximation of Newton's method.

If the exponential map is replaced by the chart $v \in T_x \mapsto (x+v)/\|x+v\| \in S^{n-1}$, Shub [1986] shows that a corresponding version of Newton's method is equivalent to the RQI.

Example 4.10 (The function $\operatorname{tr}\Theta^{\mathrm{T}}Q\Theta N$) *Let Θ, Q, $H = \operatorname{Ad}_{\Theta^{\mathrm{T}}}(Q)$, and Ω be as in Example 3.6. The second covariant differential of $f(\Theta) = \operatorname{tr}\Theta^{\mathrm{T}}Q\Theta N$ may be computed either by polarization of the second order term of $\operatorname{tr}\operatorname{Ad}_{e^{-t\Omega}}(H)N$, or by covariant differentiation of the differential $df_\Theta = -\operatorname{tr}[H,N]\Theta^{\mathrm{T}}(\cdot)$:*

$$(\nabla^2 f)_\Theta(\Theta X, \Theta Y) = -\tfrac{1}{2}\operatorname{tr}([H,\operatorname{ad}_X N] - [\operatorname{ad}_X H, N])Y,$$

where X, $Y \in \mathfrak{so}(n)$. To compute the direction $\Theta X \in T_\Theta$, $X \in \mathfrak{so}(n)$, for Newton's method, we must solve the equation $(\nabla^2 f)_\Theta(\Theta\cdot, \Theta X) = df_\Theta$, which yields the linear equation

$$L_\Theta(X) \stackrel{\mathrm{def}}{=} [H, \mathrm{ad}_X N] - [\mathrm{ad}_X H, N] = 2[H, N].$$

The linear operator $L_\Theta\colon \mathfrak{so}(n) \to \mathfrak{so}(n)$ is self-adjoint for all Θ and, in a neighborhood of the maximum, negative definite. Therefore, standard iterative techniques in the vector space $\mathfrak{so}(n)$, such as the classical conjugate gradient method, may be used to solve this equation near the maximum. The results of a numerical experiment demonstrating the convergence of Newton's method in $SO(20)$ are shown in Figure 2. As can be seen, Newton's method converged within round-off error in two iterations.

Remark 4.11 If Newton's method applied to the function $f(\Theta) = \mathrm{tr}\,\Theta^\mathsf{T} Q \Theta N$ converges to the point $\hat{\Theta}$ such that $\mathrm{Ad}_{\hat{\Theta}^\mathsf{T}}(Q) = H_\infty = \alpha N$, $\alpha \in \mathbf{R}$, then it converges cubically.

Proof By covariant differentiation of $\nabla^2 f$, the third covariant differential of f at Θ evaluated at the tangent vectors ΘX, ΘY, $\Theta Z \in T_\Theta$, X, Y, $Z \in \mathfrak{so}(n)$, is

$$(\nabla^3 f)_\Theta(\Theta X, \Theta Y, \Theta Z) = -\tfrac{1}{4}\,\mathrm{tr}\big([\mathrm{ad}_Y\,\mathrm{ad}_Z H, N] - [\mathrm{ad}_Z\,\mathrm{ad}_Y N, H]$$
$$+ [H, \mathrm{ad}_{\mathrm{ad}_Y Z} N] - [\mathrm{ad}_Y H, \mathrm{ad}_Z N] + [\mathrm{ad}_Y N, \mathrm{ad}_Z H]\big) X.$$

If $H = \alpha N$, $\alpha \in \mathbf{R}$, then $(\nabla^3 f)_\Theta(\cdot, \Theta X, \Theta X) = 0$. Therefore, the second order terms on the right hand side of Equation (4.5) vanish at the critical point. The remark follows from the smoothness of f. $\square$

This remark illuminates how rapid convergence of Newton's method applied to the function f can be achieved in some instances. If $E_{ij} \in \mathfrak{so}(n)$ is a matrix with entry $+1$ at element (i, j), -1 at element (j, i), and zero elsewhere, $X = \sum_{i<j} x^{ij} E_{ij}$, $H = \mathrm{diag}(h_1, \ldots, h_n)$, and $N = \mathrm{diag}(\nu_1, \ldots, \nu_n)$, then

$$(\nabla^3 f)_\Theta(\Theta E_{ij}, \Theta X, \Theta X) =$$
$$-2 \sum_{k \neq i,j} x^{ik} x^{jk}\big((h_i\nu_j - h_j\nu_i) + (h_j\nu_k - h_k\nu_j) + (h_k\nu_i - h_i\nu_k)\big).$$

If the h_i are close to $\alpha\nu_i$, $\alpha \in \mathbf{R}$, for all i, then $(\nabla^3 f)_\Theta(\cdot, \Theta X, \Theta X)$ may be small, yielding a fast rate of quadratic convergence.

Example 4.12 (Jacobi's method) *Let π be the projection of a square matrix onto its diagonal, and let Q be as above. Consider the maximization of the function $f(\Theta) = \mathrm{tr}\,H\pi(H)$, $H = \mathrm{Ad}_{\Theta^\mathsf{T}}(Q)$, on the special orthogonal group. This is equivalent to minimizing the sum of the squares of the off-diagonal elements of H (Golub and Van Loan [1983] derive the classical Jacobi method). The gradient of this function at Θ is $2\Theta[H, \pi(H)]$ (Chu and Driessel [1990]). By repeated covariant* *differentiation of f, we find*

$$(\nabla f)_I(X) = -2\operatorname{tr}[H, \pi(H)]X$$

$$(\nabla^2 f)_I(X, Y) = -\operatorname{tr}\big([H, \operatorname{ad}_X \pi(H)] - [\operatorname{ad}_X H, \pi(H)] - 2[H, \pi(\operatorname{ad}_X H)]\big)Y$$

$$(\nabla^3 f)_I(X, Y, Z) = -\tfrac{1}{2}\operatorname{tr}\big([\operatorname{ad}_Y \operatorname{ad}_Z H, \pi(H)] - [\operatorname{ad}_Z \operatorname{ad}_Y \pi(H), H]$$
$$+ [H, \operatorname{ad}_{\operatorname{ad}_Y Z}\, \pi(H)] - [\operatorname{ad}_Y H, \operatorname{ad}_Z \pi(H)] + [\operatorname{ad}_Y \pi(H), \operatorname{ad}_Z H]$$
$$+ 2[H, \pi(\operatorname{ad}_Y \operatorname{ad}_Z H)] + 2[H, \pi(\operatorname{ad}_Z \operatorname{ad}_Y H)]$$
$$+ 2[\operatorname{ad}_Y H, \pi(\operatorname{ad}_Z H)] - 2[H, \operatorname{ad}_Y \pi(\operatorname{ad}_Z H)]$$
$$+ 2[\operatorname{ad}_Z H, \pi(\operatorname{ad}_Y H)] - 2[H, \operatorname{ad}_Z \pi(\operatorname{ad}_Y H)]\big)X$$

where I is the identity matrix and X, Y, $Z \in \mathfrak{so}(n)$. It is easily shown that if $[H, \pi(H)] = 0$, i.e., if H is diagonal, then $(\nabla^3 f)_\Theta(\cdot, \Theta X, \Theta X) = 0$ (n.b. $\pi(\operatorname{ad}_X H) = 0$). Therefore, by the same argument as the proof of Remark 4.11, Newton's method applied to the function $\operatorname{tr} H\pi(H)$ converges cubically.

5 Conjugate gradient on Riemannian manifolds

The method of steepest descent provides an optimization technique which is relatively inexpensive per iteration, but converges relatively slowly. Each step requires the computation of a geodesic and a gradient direction. Newton's method provides a technique which is more costly both in terms of computational complexity and memory requirements, but converges relatively rapidly. Each step requires the computation of a geodesic, a gradient, a second covariant differential, and its inverse. In this section we describe the conjugate gradient method, which has the dual advantages of algorithmic simplicity and superlinear convergence.

Hestenes and Stiefel [1952] first used conjugate gradient methods to compute the solutions of linear equations, or, equivalently, to compute the minimum of a quadratic form on $\mathbf{R}^n$. This approach can be modified to yield effective algorithms to compute the minima of nonquadratic functions on $\mathbf{R}^n$. In particular, Fletcher and Reeves [1964], and Polak [1971], provide algorithms based upon the assumption that the second order Taylor expansion of the function to be minimized sufficiently approximates this function near the minimum. In addition, Davidon, Fletcher, and Reeves developed the variable metric methods (Fletcher [1987], Polak [1971]), but these will not be discussed here. One noteworthy feature of conjugate gradient algorithms on $\mathbf{R}^n$ is that when the function in question is quadratic, they compute its minimum in no more than n steps.

The conjugate gradient method on Euclidean space is uncomplicated. Given a function $f\colon \mathbf{R}^n \to \mathbf{R}$ with continuous second derivatives and a local minimum at $\hat{x}$, and an initial point $x_0 \in \mathbf{R}^n$, the algorithm is initialized by computing the (negative) gradient direction $G_0 = H_0 = -(\operatorname{grad} f)_{x_0}$. The recursive part of the algorithm involves (i) a line minimization of f along the affine space $x_i + tH_i$, $t \in \mathbf{R}$, where the minimum occurs at, say, $t = \lambda_i$, (ii) computation of the step $x_{i+1} = x_i + \lambda_i H_i$, (iii) computation of the (negative) gradient $G_{i+1} = -(\operatorname{grad} f)_{x_{i+1}}$, and (iv) computation of the next direction for line minimization,

$$H_{i+1} = G_{i+1} + \gamma_i H_i, \tag{5.1}$$

where γ_i is chosen such that H_i and H_{i+1} conjugate with respect to the Hessian matrix of f at $\hat{x}$. When f is a quadratic form represented by the symmetric positive definite matrix Q, the conjugacy condition becomes $H_i^{\mathrm{T}} Q H_{i+1} = 0$; therefore, $\gamma_i = -H_i^{\mathrm{T}} Q G_{i+1} / H_i^{\mathrm{T}} Q H_i$. It can be shown in this case that the sequence of vectors G_i are all mutually orthogonal and the sequence of vectors H_i are all mutually conjugate with respect to Q. Using these facts, the computation of γ_i may be simplified with the observation that $\gamma_i = \|G_{i+1}\|^2 / \|G_i\|^2$ (Fletcher-Reeves) or $\gamma_i = (G_{i+1} - G_i)^{\mathrm{T}} G_{i+1} / \|G_i\|^2$ (Polak-Ribière). When f is not quadratic, it is assumed that its second order Taylor expansion sufficiently approximates f in a neighborhood of the minimum, and the γ_i are chosen so that H_i and H_{i+1} are conjugate with respect to the matrix $(\partial^2 f / \partial x^i \partial x^j)(x_{i+1})$ of second partial derivatives of f at x_{i+1}. It may be desirable to "reset" the algorithm by setting $H_{i+1} = G_{i+1}$ every rth step (frequently, $r = n$) because the conjugate gradient method does not, in general, converge in n steps if the function f is nonquadratic. However, if f is closely approximated by a quadratic function, the reset strategy may be expected to converge rapidly, whereas the unmodified algorithm may not be.

Many of these ideas have straightforward generalizations in the geometry of Riemannian manifolds; several of them have already appeared. We need only make the following definition.

Definition 5.1 *Given a tensor field ω of type $(0,2)$ on M such that for p in M, $\omega_p \colon T_p \times T_p \to \mathbf{R}$ is a symmetric bilinear form, the tangent vectors X and Y in T_p are said to be ω_p-conjugate or conjugate with respect to ω_p if $\omega_p(X, Y) = 0$.*

An outline of the conjugate gradient method on Riemannian manifolds may now be given. Let M be an n-dimensional Riemannian manifold with Riemannian structure g and Levi-Civita connection ∇, and let $f \in C^\infty(M)$ have a local minimum at $\hat{p}$. As in the conjugate gradient method on Euclidean space, choose an initial point p_0 in M and compute the (negative) gradient directions $G_0 = H_0 = -(\mathrm{grad} f)_{p_0}$ in T_{p_0}. The recursive part of the algorithm involves minimizing f along the geodesic $t \mapsto \exp_{p_i} t H_i$, $t \in \mathbf{R}$, making a step along the geodesic to the minimum point $p_{i+1} = \exp \lambda_i H_i$, computing $G_{i+1} = -(\mathrm{grad} f)_{p_{i+1}}$, and computing the next direction in $T_{p_{i+1}}$ for geodesic minimization. This direction is given by the formula

$$H_{i+1} = G_{i+1} + \gamma_i \tau H_i, \tag{5.2}$$

where τ is the parallel translation with respect to the geodesic step from p_i to p_{i+1}, and γ_i is chosen such that τH_i and H_{i+1} are $(\nabla^2 f)_{p_{i+1}}$-conjugate, i.e.,

$$\gamma_i = -\frac{(\nabla^2 f)_{p_{i+1}}(\tau H_i, G_{i+1})}{(\nabla^2 f)_{p_{i+1}}(\tau H_i, \tau H_i)}. \tag{5.3}$$

Equation (5.3) is, in general, expensive to use because the second covariant differential of f appears. However, we can use the Taylor expansion of df about p_{i+1} to compute an efficient approximation of γ_i. By the fact that $p_i = \exp_{p_{i+1}}(-\lambda_i \tau H_i)$ and by Equation (4.1), we have

$$\tau df_{p_i} = \tau df_{\exp_{p_{i+1}}(-\lambda_i \tau H_i)} = df_{p_{i+1}} - \lambda_i (\nabla^2 f)_{p_{i+1}}(\cdot, \tau H_i) + \mathrm{h.o.t.}$$

Therefore, the numerator of the right hand side of Equation (5.3) multiplied by the step size λ_i can be approximated by the equation

$$\lambda_i (\nabla^2 f)_{p_{i+1}}(\tau H_i, G_{i+1}) = df_{p_{i+1}}(G_{i+1}) - (\tau df_{p_i})(G_{i+1})$$
$$= -\langle G_{i+1} - \tau G_i, G_{i+1}\rangle$$

because, by definition, $G_i = -(\operatorname{grad} f)_{p_i}$, $i = 0,\ 1,\ \dots$, and for any X in $T_{p_{i+1}}$, $(\tau df_{p_i})(X) = df_{p_i}(\tau^{-1}X) = \langle(\operatorname{grad} f)_{p_i}, \tau^{-1}X\rangle = \langle\tau(\operatorname{grad} f)_{p_i}, X\rangle$. Similarly, the denominator of the right hand side of Equation (5.3) multiplied by λ_i can be approximated by the equation

$$\lambda_i (\nabla^2 f)_{p_{i+1}}(\tau H_i, \tau H_i) = df_{p_{i+1}}(\tau H_i) - (\tau df_{p_i})(\tau H_i)$$
$$= \langle G_i, H_i\rangle$$

because $\langle G_{i+1}, \tau H_i\rangle = 0$ by the assumption that f is minimized along the geodesic $t \mapsto \exp t H_i$ at $t = \lambda_i$. Combining these two approximations with Equation (5.3), we obtain a formula for γ_i that is relatively inexpensive to compute:

$$\gamma_i = \frac{\langle G_{i+1} - \tau G_i, G_{i+1}\rangle}{\langle G_i, H_i\rangle}. \tag{5.4}$$

Of course, as the connection ∇ is compatible with the metric g, the denominator of Equation (5.4) may be replaced, if desired, by $\langle\tau G_i, \tau H_i\rangle$.

The conjugate gradient method may now be presented in full.

Algorithm 5.2 (Conjugate gradient method) *Let M be a complete Riemannian manifold with Riemannian structure g and Levi-Civita connection ∇, and let f be a C^∞ function on M.*

    Step 0: *Select $p_0 \in M$, compute $G_0 = H_0 = -(\operatorname{grad} f)_{p_0}$, and set $i = 0$.*

    Step 1: *Compute λ_i such that*

$$f(\exp_{p_i}\lambda_i H_i) \leq f(\exp_{p_i}\lambda H_i)$$

for all $\lambda \geq 0$.

    Step 2: *Set $p_{i+1} = \exp_{p_i}\lambda_i H_i$.*

    Step 3: *Set*

$$G_{i+1} = -(\operatorname{grad} f)_{p_{i+1}},$$
$$H_{i+1} = G_{i+1} + \gamma_i \tau H_i, \qquad \gamma_i = \frac{\langle G_{i+1} - \tau G_i, G_{i+1}\rangle}{\langle G_i, H_i\rangle},$$

where τ is the parallel translation with respect to the geodesic from p_i to p_{i+1}. If $i \equiv n - 1 \pmod{n}$, set $H_{i+1} = G_{i+1}$. Increment i, and go to Step 1.

    Theorem 5.3 *Let $f \in C^\infty(M)$ have a nondegenerate critical point at $\hat{p}$ such that the Hessian $(d^2 f)_{\hat{p}}$ is positive definite. Let p_i be a sequence of points in M generated by Algorithm 5.2 converging to $\hat{p}$. Then there exists a constant $\theta > 0$ and an integer N such that for all $i \geq N$,*

$$d(p_{i+n}, \hat{p}) \leq \theta d^2(p_i, \hat{p}).$$

Note that linear convergence is already guaranteed by Theorem 3.3.

Proof If $p_j = \hat{p}$ for some integer j, the assertion becomes trivial; assume otherwise. Recall that if $X_1, \ldots, X_n$ is some basis for $T_{\hat{p}}$, then the map $\exp_{\hat{p}}(a^1 X_1 + \cdots + a^n X_n) \xrightarrow{\nu} (a^1, \ldots, a^n)$ defines a set of normal coordinates at $\hat{p}$. Let $N_{\hat{p}}$ be a normal neighborhood of $\hat{p}$ on which the normal coordinates $\nu = (x^1, \ldots, x^n)$ are defined. Consider the map $\nu_* f \stackrel{\text{def}}{=} f \circ \nu^{-1} \colon \mathbf{R}^n \to \mathbf{R}$. By the smoothness of f and $\exp$, $\nu_* f$ has a critical point at $0 \in \mathbf{R}^n$ such that the Hessian matrix of $\nu_* f$ at 0 is positive definite. Indeed, by the fact that $(d\exp)_0 = \mathrm{id}$, the ijth component of the Hessian matrix of $\nu_* f$ at 0 is given by $(d^2 f)_{\hat{p}}(X_i, X_j)$.

Therefore, there exists a neighborhood U of $0 \in \mathbf{R}^n$, a constant $\theta' > 0$, and an integer N, such that for any initial point $x_0 \in U$, the conjugate gradient method on Euclidean space (with resets) applied to the function $\nu_* f$ yields a sequence of points x_i converging to 0 such that for all $i \geq N$,

$$\|x_{i+n}\| \leq \theta' \|x_i\|^2.$$

See Polak [1971], p. 260ff for a proof of this fact. Let $x_0 = \nu(p_0)$ in U be an initial point. Because $\exp$ is not an isometry, Algorithm 5.2 yields a different sequence of points in $\mathbf{R}^n$ than the classical conjugate gradient method on $\mathbf{R}^n$ (upon equating points in a neighborhood of $\hat{p} \in M$ with points in a neighborhood of $0 \in \mathbf{R}^n$ via the normal coordinates).

Nevertheless, the amount by which $\exp$ fails to preserve inner products can be quantified via the Gauss Lemma and Jacobi's equation; see, for example, Cheeger and Ebin [1975], or the appendices of Karcher [1977]. Let t be small, and let $X \in T_{\hat{p}}$ and $Y \in T_{tX}(T_{\hat{p}}) \cong T_{\hat{p}}$ be orthonormal tangent vectors. The amount by which the exponential map changes the length of tangent vectors is approximated by the Taylor expansion

$$\|d\exp(tY)\|^2 = t^2 - \tfrac{1}{3} K t^4 + \text{h.o.t.}$$

where K is the sectional curvature of M along the section in $T_{\hat{p}}$ spanned by X and Y. Therefore, near $\hat{p}$ Algorithm 5.2 differs from the conjugate gradient method on $\mathbf{R}^n$ applied to the function $\nu_* f$ only by third order and higher terms. Thus both algorithms have the same rate of convergence. The theorem follows. $\square$

Example 5.4 (Rayleigh's quotient on the sphere) *Applied to Rayleigh's quotient on the sphere, the conjugate gradient method provides an efficient technique to compute the eigenvectors corresponding to the largest or smallest eigenvalue of a real symmetric matrix. Let S^{n-1} and $\rho(x) = x^{\mathrm{T}} Q x$ be as in Examples 3.5 and 4.6. From Algorithm 5.2, we have the following algorithm.*

Algorithm 5.5 (CG method for the extreme eigenvalue/eigenvector) *Let Q be a real symmetric n-by-n matrix.*

    Step 0: *Select x_0 in $\mathbf{R}^n$ such that $x_0^{\mathrm{T}} x_0 = 1$, compute $G_0 = H_0 = (Q - \rho(x_0)I)x_0$, and set $i = 0$.*

    Step 1: *Compute c, s, and $v = 1 - c = s^2/(1 + c)$, such that $\rho(x_i c + h_i s)$ is maximized, where $c^2 + s^2 = 1$ and $h_i = H_i/\|H_i\|$. This can be accomplished*

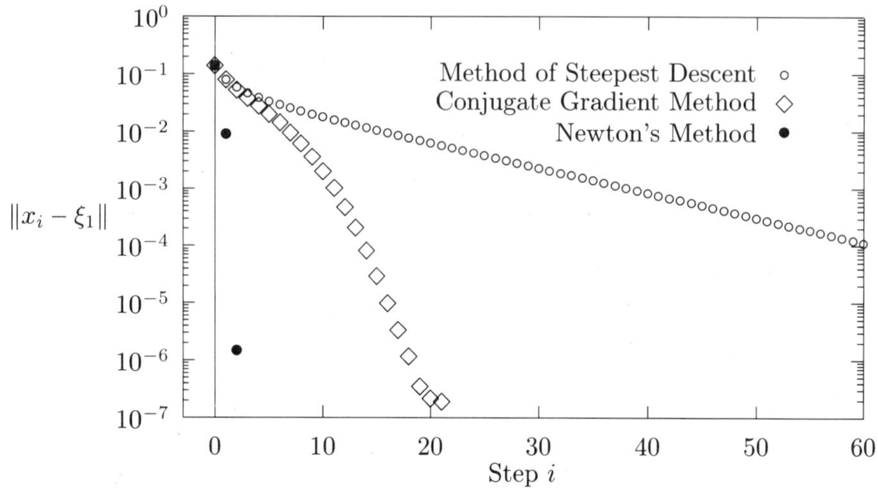

Figure 1 Maximization of Rayleigh's quotient $x^{\mathrm{T}}Qx$ on $S^{20} \subset \mathbf{R}^{21}$, where $Q = \mathrm{diag}(21, \ldots, 1)$. The ith iterate is x_i, and ξ_1 is the eigenvector corresponding to the largest eigenvalue of Q. Algorithm 4.7 was used for Newton's method and Algorithm 5.5 was used for the conjugate gradient method.

by geodesic minimization, or by the formulae

$$c = \left(\tfrac{1}{2}(1 + b/r)\right)^{\frac{1}{2}} \quad \text{if } b \geq 0, \quad \text{or} \quad s = \left(\tfrac{1}{2}(1 - b/r)\right)^{\frac{1}{2}} \quad \text{if } b \leq 0,$$
$$s = a/(2rc) \qquad\qquad\qquad\qquad c = a/(2rs)$$

where $a = 2x_i^{\mathrm{T}}Qh_i$, $b = x_i^{\mathrm{T}}Qx_i - h_i^{\mathrm{T}}Qh_i$, *and* $r = \sqrt{(a^2 + b^2)}$.

Step 2: *Set*

$$x_{i+1} = x_i c + h_i s, \quad \tau H_i = H_i c - x_i \|H_i\| s, \quad \tau G_i = G_i - (h_i^{\mathrm{T}} G_i)(x_i s + h_i v).$$

Step 3: *Set*

$$G_{i+1} = \left(Q - \rho(x_{i+1})I\right)x_{i+1},$$
$$H_{i+1} = G_{i+1} + \gamma_i \tau H_i, \qquad \gamma_i = \frac{(G_{i+1} - \tau G_i)^{\mathrm{T}} G_{i+1}}{G_i^{\mathrm{T}} H_i}.$$

If $i \equiv n - 1 \pmod{n}$, *set* $H_{i+1} = G_{i+1}$. *Increment* i, *and go to Step 1.*

The convergence rate of this algorithm to the eigenvector corresponding to the largest eigenvalue of Q is given by Theorem 5.3. This algorithm costs one matrix-vector multiplication (relatively inexpensive when Q is sparse), one geodesic minimization or computation of $\rho(h_i)$, and $10n$ flops per iteration. The results of a numerical experiment demonstrating the convergence of Algorithm 5.5 on S^{20} are shown in Figure 1.

Fuhrmann and Liu [1984] provide a conjugate gradient algorithm for Rayleigh's quotient on the sphere that uses an azimuthal projection onto tangent planes.

Example 5.6 (The function $\mathrm{tr}\,\Theta^{\mathrm{T}}Q\Theta N$) *Let* Θ, Q, *and* H *be as in Examples 3.6 and 4.10. As before, the natural Riemannian structure of $SO(n)$ is used. Let* X, $Y \in \mathfrak{so}(n)$. *The parallel translation of Y along the geodesic e^{tX} is given by*

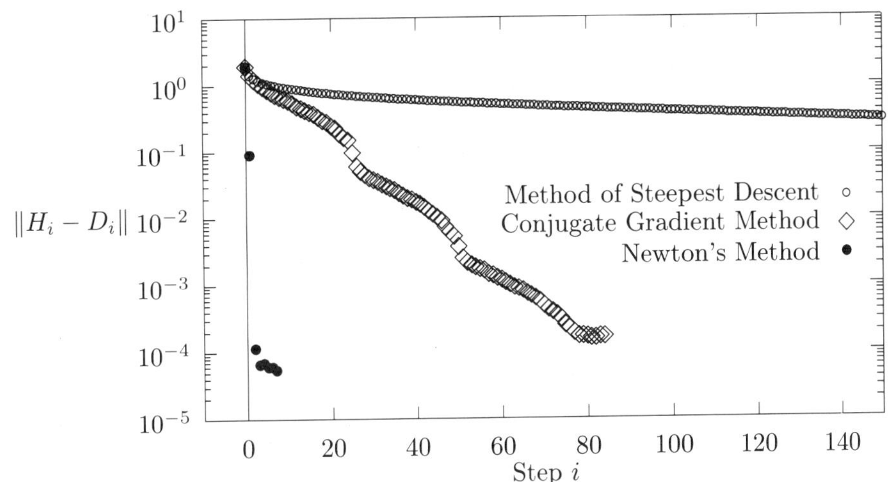

Figure 2 Maximization of $\operatorname{tr}\Theta^{\mathrm{T}}Q\Theta N$ on $SO(20)$ (dimension $SO(20) =$ 190), where $N = \operatorname{diag}(20,\dots,1)$. The ith iterate is $H_i = \Theta_i^{\mathrm{T}}Q\Theta_i$, D_i is the diagonal matrix of eigenvalues of H_i, H_0 is near N, and $\|\cdot\|$ is the norm induced by the standard inner product on $\mathfrak{gl}(n)$. Geodesics and parallel translation were computed using the algorithm of Ward and Gray [1978a] and Ward and Gray [1978b]; the step sizes for the method of steepest descent and the conjugate gradient method were computed using Brockett's estimate (Brockett, (to appear)).

the formula $\tau Y = L_{e^{tX}*}e^{-(t/2)X}Ye^{(t/2)X}$, where L_g denotes left translation by g. Brockett's estimate (n.b. Equation (3.13)) for the step size may be used in Step 1 of Algorithm 5.2. The results of a numerical experiment demonstrating the convergence of the conjugate gradient method in $SO(20)$ are shown in Figure 2.

References

Bertsekas, D.P. [1982a] *Projected Newton methods for optimization problems with simple constraints*, SIAM J. Cont. Opt., **20**, 221–246.

______[1982b] *Constrained Optimization and Lagrange Multiplier Methods*, New York, Academic Press.

Bloch, A.M., Brockett, R.W., and Ratiu, T.S. [1990] *A new formulation of the generalized Toda lattice equations and their fixed point analysis via the momentum map*, Bull. Amer. Math. Soc., **23** (2), 477–485.

______[1992] *Completely integrable gradient flows*, Commun. Math. Phys., **147**, 57–74.

Botsaris, C.A. [1978a] *Differential gradient methods*, J. Math. Anal. Appl., **63**, 177–198.

______[1978b] *A class of differential descent methods for constrained optimization*, J. Math. Anal. Appl., **79**, 96–112.

______[1981] *Constrained optimization along geodesics*, J. Math. Anal. Appl., **79**, 295–306.

Brockett, R.W. [1989] *Least squares matching problems*, Lin. Alg. Appl. **122, 123, 124**, 761–777.

—— [1991] *Dynamical systems that sort lists, diagonalize matrices, and solve linear programming problems*, Lin. Alg. Appl., **146**, 79–91.

—— [1993] *Differential geometry and the design of gradient algorithms*, Proc. Symp. Pure Math. (R. Green and S.T. Yau, eds.), Providence, RI, Amer. Math. Soc., Vol. 54, Part 1, pp. 69-92.

Brown, A.A., and Bartholomew-Biggs, M.C. [1989] *Some effective methods for unconstrained optimization based on the solution of systems of ordinary differential equations*, J. Optim. Theory Appl., **62** (2), 211–224.

Cheeger, J., and Ebin, D.G. [1975] *Comparison Theorems in Riemannian Geometry*, Amsterdam, North-Holland Publishing Company.

Chu, M.T. [1986] *Curves on S^{n-1} that lead to eigenvalues or their means of a matrix*, SIAM J. Alg. Disc. Meth., **7** (3), 425–432.

Chu, M.T., and Driessel, K. [1990] *The projected gradient method for least squares matrix approximations with spectral constraints*, SIAM J. Numer. Anal., **27** (4), 1050–1060.

Dunn, J.C. [1990] *Newton's method and the Goldstein step length rule for constrained minimization problems*, SIAM J. Cont. Opt., **18**, 659–674.

—— [1981] *Global and asymptotic convergence rate estimates for a class of projected gradient processes*, SIAM J. Cont. Opt., **19**, 368–400.

Faybusovich, L. [1991] *Hamiltonian structure of dynamical systems which solve linear programming problems*, Phys. D, **53**, 217–232.

Fletcher, R. [1987] *Practical Methods of Optimization*, 2d ed., New York, Wiley & Sons.

Fletcher, R., and Reeves, C.M. *Function minimization by conjugate gradients*, Comput. J., **7** (2), 149–154.

Fuhrmann, D. R., and Liu, B. [1984] *An iterative algorithm for locating the minimal eigenvector of a symmetric matrix*, Proc. IEEE ICASSP 84, pp. 45.8.1–4.

Gill, P.E., and Murray, W.[1974] *Newton-type methods for linearly constrained optimization*; in Numerical Methods for Constrained Optimization (P. E. Gill and W. Murray, eds.), London, Academic Press.

Golub, G.H., and Van Loan, C. [1983] *Matrix Computations*. Baltimore, MD, Johns Hopkins University Press.

Golubitsky, M., and Guillemin, V. [1973] *Stable Mappings and Their Singularities*. New York, Springer-Verlag.

Helgason, S. [1978] *Differential Geometry, Lie Groups, and Symmetric Spaces*. New York, Academic Press.

Helmke, U. [1991] *Isospectral flows on symmetric matrices and the Riccati equation*, Systems & Control Lett., **16**, 159–165.

Hestenes, M.R., and Stiefel, E. [1952] *Methods of conjugate gradients for solving linear systems*, J. Res. Nat. Bur. Stand., **49**, 409–436.

Hirsch, M.W., and Smale, S. [1979] *On algorithms for solving $f(x) = 0$*, Comm. Pure Appl. Math., **32**, 281–312.

Karcher, H. [1977] *Riemannian center of mass and mollifier smoothing*, Comm. Pure Appl. Math., **30**, 509–541.

Kobayashi, S., and Nomizu, K. [1969] *Foundations of Differential Geometry*, **2**, New York, Wiley Interscience Publishers.

Lagarias, J.C. [1991] *Monotonicity properties of the Toda flow, the QR-flow, and subspace iteration*, SIAM J. Numer. Anal. Appl., **12** (3), 449–462.

Luenberger, D.G. [1973] *Introduction to Linear and Nonlinear Programming*, Reading, MA, Addison-Wesley.

Moler, C., and Van Loan, C. [1978] *Nineteen dubious ways to compute the exponential of a matrix*, SIAM Rev., **20** (4), 801–836.

Nomizu, K. [1954] *Invariant affine connections on homogeneous spaces.* Amer. J. Math., **76**, 33–65.

Parlett, B. [1980] *The Symmetric Eigenvalue Problem*, Englewood Cliffs, NJ, Prentice-Hall.

Perkins, J.E., Helmke, U., and Moore, J.B. [1990] *Balanced realizations via gradient flow techniques*, Systems & Control Lett., **14**, 369–380.

Polak, E. [1971] *Computational Methods in Optimization.* New York, Academic Press.

Rudin, W. [1976] *Principles of Mathematical Analysis.* 3d ed., New York, McGraw-Hill.

Sargent, R.W.H. [1974] *Reduced gradient and projection methods for nonlinear programming*; in Numerical Methods for Constrained Optimization (P.E. Gill and W. Murray, eds.), London, Academic Press.

Shub, M. [1986] *Some remarks on dynamical systems and numerical analysis*, in Dynamical Systems and Partial Differential Equations, Proc. VII ELAM., (L. Lara-Carrero and J. Lewowicz, eds.) Caracas, Equinoccio, U. Simón Bolívar, pp. 69–92.

Shub, M., and Smale, S. [1985] *Computational complexity: On the geometry of polynomials and a theory of cost*, Part I, Ann. Scient. Éc. Norm. Sup., **4** (18), 107–142.

———[1986a] *Computational complexity: On the geometry of polynomials and a theory of cost*, Part II, SIAM J. Comput., **15** (1), 145–161.

———[1986b] *On the existence of generally convergent algorithms*, J. Complex., **2**, 2–11.

Smale, S. [1981] *The fundamental theorem of algebra and computational complexity*, Bull. Amer. Math. Soc., **4** (1), 1–36.

———[1985] *On the efficiency of algorithms in analysis*, Bull. Amer. Math. Soc., **13** (2), 87–121.

Smith, S.T. [1991] *Dynamical systems that perform the singular value decomposition*, Systems & Control Lett., **16**, 319–327.

Spivak, M. [1979] *A Comprehensive Introduction to Differential Geometry*, 2nd ed. **1, 2, 5**, Houston, TX, Publish or Perish, Inc.

Ward, R.C., and Gray, L.J. [1978a] *Eigensystem computation for skew-symmetric matrices and a class of symmetric matrices*, ACM Trans. Math. Softw., **4** (3), 278–285.

Ward, R.C., and Gray, L.J. [1978b] *Algorithm 530: An algorithm for computing the eigensystem of skew-symmetric matrices and a class of symmetric matrices*, ACM Trans. Math. Softw., **4** (3), 286–289. See also *Collected Algorithms from ACM*, **3**, New York, Assoc. Comput. Mach.

Fields Institute Communications
Volume **3**, 1994

On the Number of Real Roots
of a Sparse Polynomial System

Bernd Sturmfels
Department of Mathematics
Cornell University
Ithaca, New York, USA
14853

Abstract. We estimate the asymptotic number of real roots of a system of polynomial equations under a multiplicative deformation which preserves the Newton polytope.

1 Introduction

We consider a system of d polynomial equations in d variables $\mathbf{x} = (x_1, \ldots, x_d)$:

$$
\begin{aligned}
f_1(\mathbf{x}) &= c_{11}\mathbf{x}^{a_1} + c_{12}\mathbf{x}^{a_2} + \ldots + c_{1n}\mathbf{x}^{a_n} &= 0 \\
\ldots & \qquad \ldots \quad \ldots \quad \ldots \quad \ldots \quad \ldots & \\
f_d(\mathbf{x}) &= c_{d1}\mathbf{x}^{a_1} + c_{d2}\mathbf{x}^{a_2} + \ldots + c_{dn}\mathbf{x}^{a_n} &= 0
\end{aligned}
\qquad (1.1)
$$

where $\mathcal{A} = \{a_1, \ldots, a_n\} \subset \mathbf{N}^d$, $\mathbf{x}^{a_i} = x_1^{a_{i1}} x_2^{a_{i2}} \cdots x_d^{a_{id}}$, and the coefficients c_{ij} are real numbers. The convex polytope $Q = conv(\mathcal{A})$ in $\mathbf{R}^d$ is called the *Newton polytope* of the system (1.1). It has the *normalized volume* $v(\mathcal{A}) := vol(Q) \cdot d\,!$.

Theorem 1.1 *(Koushnirenko [1975]) For all coefficient matrices (c_{ij}) in a dense subset $\mathcal{U}_\mathcal{A}$ of $\mathbf{R}^{d \times n}$, the system (1.1) has $v(\mathcal{A})$ distinct zeros in the complex torus $(\mathbf{C}^*)^d$.*

Bernstein [1975] gives an extension of Theorem 1.1 to *mixed systems*, where each polynomial $f_i(\mathbf{x})$ has a different set of exponent vectors $\mathcal{A}_i \subset \mathbf{N}^d$. Here we restrict ourselves to the unmixed case, and we assume $dim(Q) = d$, or, equivalently, $v(\mathcal{A}) > 0$. All our results can be extended to mixed systems using the techniques in Pedersen and Sturmfels [1993], Sect. 7. The dense subset $\mathcal{U}_\mathcal{A} \subset \mathbf{R}^{d \times n}$ appearing in Theorem 1.1 is Zariski-open, i.e., $\mathcal{U}_\mathcal{A}$ is the set of non-zeros of a finite system of

1991 *Mathematics Subject Classification.* Primary 12D10; Secondary 65A05.

Supported in part by the National Science Foundation and the A.P. Sloan Foundation.

Supported in part by the Ministry of Colleges and Universities of Ontario and the Natural Sciences and Engineering Research Council of Canada while visiting The Fields Institute.

polynomials. This system consists of the face resultants appearing in Pedersen and Sturmfels [1993], Theorem 3.1, plus a certain discriminant which rules out multiple roots.

This note is concerned with following question: *How many of the $v(\mathcal{A})$ complex zeros are* <u>real</u> *zeros?* Let $\rho(\mathcal{A})$ denote the maximum number of real roots $\mathbf{x} \in (\mathbf{R}^*)^d$ of (1.1) as (c_{ij}) ranges over $\mathcal{U}_\mathcal{A}$. Similarly we write $\rho_+(\mathcal{A})$ for the maximum number of real positive roots $\mathbf{x}$ in $(\mathbf{R}_+)^d$. Counting the 2^d orthants in $\mathbf{R}^d$, we get the obvious inequality

$$\rho(\mathcal{A}) \quad \le \quad 2^d \cdot \rho_+(\mathcal{A}). \tag{1.2}$$

The theory of *Fewnomials*, due to Khovanskii [1991], shows that the number of real roots is generally much smaller than the number of complex roots. There exists an upper bound for $\rho(\mathcal{A})$ which depends only on the dimension d and the number of terms $n = |\mathcal{A}|$.

Theorem 1.2 *(Khovanskii [1980], [1991]) The number of positive real roots of (1.1) satisfies*

$$\rho_+(\mathcal{A}) \quad \le \quad 2^{n(n-1)/2} \cdot (d+1)^n. \tag{1.3}$$

This theorem raises the following natural question.

Problem 1.3 *Find lower bounds and more precise upper bounds for $\rho(\mathcal{A})$, the maximum number of real roots, in terms of the combinatorial structure of the configuration $\mathcal{A}$.*

Very little is known at present, as is witnessed by a challenging little example.

Example 1.4 *What is the maximum number of real roots of a bivariate system with five terms?* In other words, determine the maximum $\rho(2,5)$ of the integers $\rho(\mathcal{A})$, where $\mathcal{A}$ runs over all five element subsets of $\mathbf{N}^2$.

Let $\mathcal{A} = \{(0,0),(2,0),(4,0),(0,2),(0,4)\}$. This configuration is the support of

$$(x^2 - 4)^2 + (y^2 - 3)^2 - 13 \quad = \quad (x^2 - 4)^2 - (y^2 - 3)^2 - 5 \quad = \quad 0. \tag{1.4}$$

The Newton polytope $Q = conv(\mathcal{A})$ is a triangle with area 8, therefore $v(\mathcal{A}) = 2 \cdot 8 = 16 \ge \rho(\mathcal{A})$. The specific system (1.4) has all 16 roots real, and therefore $\rho(\mathcal{A}) = 16$.

The resulting inequality $\rho(2,5) \ge 16$ is the best lower bound for $\rho(2,5)$ known to me at present. There is an embarrassingly wide gap to the upper bound from Khovanskii's Theorem 1.2. Combining (1.2) and (1.3), we have $\rho(2,5) \le 2^2 \cdot 2^{10} \cdot 3^5 = 995,328$. $\square$

In this note we venture a first step in the attack on Problem 1.3. We fix $\omega = (\omega_1, \dots, \omega_n) \in \mathbf{N}^n$, and we consider the perturbed system

$$f_1(t; x_1, \dots, x_n) \quad = \quad c_{11} t^{\omega_1} \mathbf{x}^{a_1} + c_{12} t^{\omega_2} \mathbf{x}^{a_2} + \dots + c_{1n} t^{\omega_n} \mathbf{x}^{a_n} \quad = \quad 0$$

$$\dots \qquad \dots \qquad \qquad \dots \qquad \dots \qquad \dots \qquad \dots \qquad \dots$$

$$f_d(t; x_1, \dots, x_n) \quad = \quad c_{d1} t^{\omega_1} \mathbf{x}^{a_1} + c_{d2} t^{\omega_2} \mathbf{x}^{a_2} + \dots + c_{dn} t^{\omega_n} \mathbf{x}^{a_n} \quad = \quad 0 \tag{1.5}$$

as the real parameter t tends to zero.

Our main result (Theorem 2.2) gives precise bounds for the number of real roots of (1.5) for sufficiently small $t > 0$. As a corollary we get a lower bound for $\rho(\mathcal{A})$. These bounds are stated in terms of *regular triangulations*. This concept is reviewed in the beginning of Section 2. The proof of our result is given in Section 3.

2 Bounds in terms of regular triangulations

A *subdivision* of the pair $(Q, \mathcal{A})$ is a collection $\Delta = \{\mathcal{A}_1, \ldots, \mathcal{A}_m\}$ of subsets of $\mathcal{A}$ such that

- each polytope $Q_i = conv(\mathcal{A}_i)$ has the full dimension d for $i = 1, \ldots, m$,
- $Q = Q_1 \cup Q_2 \cup \ldots \cup Q_m$, and
- each intersection $Q_i \cap Q_j$ is a common proper face of Q_i and Q_j, for $1 \leq i < j \leq m$.

The subsets $\mathcal{A}_i$ are the *cells* of the subdivision Δ. We say that Δ is a *triangulation* of $(Q, \mathcal{A})$ if each cell $\mathcal{A}_i$ has cardinality $d + 1$. An important subclass of triangulations is the class of *regular triangulations* (see e.g. Billera, Filliman and Sturmfels [1990], Gel'fand, Kapranov and Zelevinsky [1990], and Lee [1991]). In what follows we give an algebraic definition of regular triangulation, based on the results in Sturmfels [1991]. Let $y_1, \ldots, y_n, z, x_1, \ldots, x_d$ be variables, and consider the **Z**-algebra homomorphism

$$\mathbf{Z}[y_1, y_2, \ldots, y_n] \;\rightarrow\; \mathbf{Z}[x_1, \ldots, x_d, z]\,, \quad y_i \;\mapsto\; z \cdot x_1^{a_{i1}} x_2^{a_{i2}} \cdots x_d^{a_{id}}, \qquad (2.1)$$

with $a_i = (a_{i1}, \ldots, a_{id})$ as before. We define the *toric ideal* $I_{\mathcal{A}}$ to be the kernel of map (2.1). The toric ideal $I_{\mathcal{A}}$ is generated by all homogeneous polynomials of the form

$$y_1^{\mu_1} y_2^{\mu_2} \cdots y_n^{\mu_n} \;-\; y_1^{\nu_1} y_2^{\nu_2} \cdots y_n^{\nu_n}, \qquad (2.2)$$

where $\mu_1 a_1 + \ldots + \mu_n a_n = \nu_1 a_1 + \ldots + \nu_n a_n$ and $\mu_1 + \ldots + \mu_n = \nu_1 + \ldots + \nu_n$. The relations (2.2) correspond to affine dependencies of $\mathcal{A}$.

We fix $\omega \in \mathbf{N}^n$. For each polynomial $f \in \mathbf{Z}[y_1, \ldots, y_n]$ we consider $f(t^{\omega_1} y_1, \ldots, t^{\omega_n} y_n)$ as a univariate polynomial in the parameter t. Its leading coefficient $init_\omega(f)$ is a polynomial in $\mathbf{Z}[x_1, \ldots, x_n]$, called the *initial form* of f with respect to ω. We define the *initial ideal* of $I_{\mathcal{A}}$ as

$$init_\omega(I_{\mathcal{A}}) \quad := \quad \langle\, init_\omega(f) \; : \; f \in I_{\mathcal{A}} \,\rangle. \qquad (2.3)$$

For sufficiently generic ω the initial ideal $init_\omega(I_{\mathcal{A}})$ is generated by monomials. We now suppose that this is the case. In the language of Gröbner basis theory: the vector ω represents a *term order* for $I_{\mathcal{A}}$. A monomial $y_1^{\mu_1} \cdots y_n^{\mu_n}$ is said to be *standard* if $y_1^{\mu_1} \cdots y_n^{\mu_n}$ does not lie in $I_{\mathcal{A}}$. Let Δ_ω denote the collection of all $(d+1)$-subsets $\{a_{i_0}, a_{i_1}, \ldots, a_{i_d}\}$ of $\mathcal{A}$ for which all powers of the square-free monomial $y_{i_0} y_{i_1} \cdots y_{i_d}$ are standard.

Theorem 2.1 *(Sturmfels [1991]) The set Δ_ω is a triangulation of $(Q, \mathcal{A})$.*

A triangulation Δ of $(Q_{\mathcal{A}})$ is called *regular* if $\Delta = \Delta_\omega$, for some $\omega \in \mathbf{N}^n$. For examples of non-regular triangulations see Billera, Filliman and Sturmfels [1990], Figure 1 and Lee [1991], Figures 2 and 6.

Fix a term order $\omega \in \mathbf{N}^n$, and let Δ_ω be the corresponding regular triangulation. Each cell $\mathcal{A}_i \in \Delta_\omega$ generates an affine lattice $\mathbf{Z}\{\mathcal{A}_i\}$ of rank d. The quotient of lattices $\mathbf{Z}^d / \mathbf{Z}\{\mathcal{A}_i\}$ is a finite abelian group of order $v(\mathcal{A}_i)$. By standard results on finite abelian groups (Hungerford [1974], Cor. 2.7), there exist unique positive integers $m_1, m_2, \ldots, m_d$, called *invariant factors*, such that m_{i-1} divides m_i for $i = 2$, $\ldots, n$, and

$$\mathbf{Z}^d / \mathbf{Z}\{\mathcal{A}_i\} \quad \simeq \quad \mathbf{Z}/m_1\mathbf{Z} \oplus \mathbf{Z}/m_2\mathbf{Z} \oplus \ldots \oplus \mathbf{Z}/m_d\mathbf{Z}. \tag{2.4}$$

We have $m_1 m_2 \cdots m_d = v(\mathcal{A}_i)$. Let $even(\mathcal{A}_i)$ denote the number of invariant factors m_i which are even. If the integer $v(\mathcal{A}_i)$ is odd, or equivalently, if $even(\mathcal{A}_i) = 0$, then we call $\mathcal{A}_i$ an *odd cell*. We now state the new result of this note.

Theorem 2.2 *Let* $(c_{ij}) \in \mathcal{U}_A$, *and let* ρ *be the number of real roots* $\mathbf{x} \in (\mathbf{R}^*)^d$ *of the system (1.5) for sufficiently small* $t > 0$. *This number satisfies the inequalities*

$$\# \text{ odd cells in } \Delta_\omega \quad \leq \quad \rho \quad \leq \quad \sum_{\mathcal{A}_i \in \Delta_\omega} 2^{even(\mathcal{A}_i)}. \tag{2.5}$$

This theorem has the following two corollaries.

Corollary 2.3 *For any configuration* $\mathcal{A} \subset \mathbf{N}^d$, *the number* $\rho(\mathcal{A})$ *is bounded below by the number of odd cells in any regular triangulation of* $(Q, \mathcal{A})$.

Corollary 2.4 *Fix* $(c_{ij}) \in \mathcal{U}_A$ *and let* Δ_ω *and* ρ *as in Theorem 2.2.*
(a) *If each cell in* Δ_ω *is odd, then the lower and upper bounds in (2.5) agree and* ρ *coincides with the total number of cells in* Δ_ω.
(b) *If each cell* $\mathcal{A}_i \in \Delta_\omega$ *has unit volume* $v(\mathcal{A}_i) = 1$, *then all complex roots of (1.5) are real for* $t \to 0$, *and we have* $\rho = \rho(\mathcal{A}) = v(\mathcal{A})$.

We illustrate these results for three classes of examples.

Example 2.5 (Dense univariate polynomials) Let $c_0, c_1, \ldots, c_n$ be fixed real numbers, let $\omega = (\omega_0, \ldots, \omega_n) \in \mathbf{N}^{n+1}$ such that $\omega_{i-1} + \omega_{i+1} \neq 2\omega_i$ for $i = 1, \ldots, n-1$, and let $t > 0$ be sufficiently small. We are interested in the number ρ of real roots of the polynomial

$$f_t(x) \quad := \quad c_0 t^{\omega_0} + c_1 t^{\omega_1} x + c_2 t^{\omega_2} x^2 + \ldots + c_n t^{\omega_n} x^n. \tag{2.6}$$

Let $(i_1, \omega_{i_1}), (i_2, \omega_{i_2}), \ldots (i_k, \omega_{i_k})$ be the vertices of the polygon $conv\{ (i, \omega_i), (i, 0) : i = 0, 1, \ldots, n\}$. We may suppose $0 = i_1 < i_2 < \ldots < i_k = n$. Let f_{odd} denote the number of differences $i_{j+1} - i_j$ which are odd. Then $f_{odd} \leq \rho \leq 2(k-1) - f_{odd}$.

If all differences $i_{j+1} - i_j$ are odd, then $f_{odd} = \rho$ for all choices of coefficients $c_i \in \mathbf{R}^*$. For instance, choose arbitrary non-zero real numbers $c_0, c_1, \ldots, c_7$, and consider

$$f_t(x) = c_0 t^3 + c_1 t^4 x + c_2 t^3 x^2 + c_3 t^1 x^3 + c_4 t^1 x^4 + c_5 t^3 x^5 + c_6 t^2 x^6 + c_7 t^2 x^7. \tag{2.7}$$

For small parameter values $t > 0$, the polynomial $f_t(x)$ has precisely $f_{odd} = 3$ real roots.

Example 2.6 (The cross-polytope) Consider the system of polynomial equations

$$\sum_{j=0}^{n} c_{ij} \prod_{k=j+1}^{n} x_k + \sum_{j=0}^{n} d_{ij} x_0 x_1 \cdots x_n \prod_{k=1}^{j} x_k \;=\; 0 \qquad (i = 1, 2, \ldots, n+1) \tag{2.8}$$

where the c_{ij}, d_{ij} are real coefficients, and the product over the empty set is defined to be 1. The corresponding set $\mathcal{A} \subset \mathbf{N}^{n+1}$ is unimodularly equivalent to the vertices of the regular $(n+1)$-dimensional cross-polytope. For $n = 2$ the cross-polytope $Q = conv(\mathcal{A})$ is the octahedron. There is a canonical regular triangulation of $\mathcal{A}$

into 2^n simplices of unit volume. The number of complex roots of (2.8) equals 2^n, and, by Corollary 2.8, this number coincides with the maximum number $\rho(\mathcal{A})$ of real roots.

Example 2.7 (Bivariate systems) Fix $d = 2$. We write $f(\Delta_\omega)$ for the number of *triangles* (= cells) in the regular triangulation Δ_ω of $\mathcal{A} = \{a_1, \dots, a_n\} \subset \mathbf{N}^2$. An easy count of triangles in planar graphs shows that $f(\Delta_\omega) \leq 2n - 5$. Theorem 2.2 implies

$$\rho \quad \leq \quad 2^2 \cdot f(\Delta_\omega) \quad \leq \quad 8n - 20.$$

Thus for $d = 2$ fixed, the number of real roots of (1.5) for $t \to 0$ is bounded above by a linear polynomial in n. Note that the Khovanskii upper bound (1.3) is exponential in n.

It is easy to see that the number $\rho(\mathcal{A})$ is bounded below by a quadratic polynomial in n. Generalizing Example 1.4, we let $f(x)$ and $g(y)$ be univariate polynomials of degree n, each having n distinct positive roots. Then the bivariate system $f(x) + g(y) = f(x) - g(y) = 0$ has n^2 real roots, but its support set $\mathcal{A}$ has only cardinality $2n + 1$. This shows that in general the number ρ in Theorem 2.2 grows much slower than $\rho(\mathcal{A})$. But how much?

We formulate this as a problem of asymptotic complexity. Let $\rho(d, n)$ denote the maximum of the numbers $\rho(\mathcal{A})$ where $\mathcal{A}$ runs over all n-element subsets of $\mathbf{N}^d$.

Problem 2.8 *For fixed $d \geq 2$, does $\rho(d, n)$ grow polynomially or exponentially in n ?*

3 The proof

Our proof of Theorem 2.2 is algorithmic. We describe an explicit procedure for computing the ρ real roots of (1.5) for $t \to 0$. We start with a "subroutine" for the base case $n = d + 1$.

Proof of Theorem 2.2 (Part I: $n = d + 1$). We show that (2.5) holds for the number of roots of (1.1), for all $(c_{ij}) \in \mathcal{U}_\mathcal{A}$. Using elementary row operations, we transform the $d \times (d + 1)$-coefficient matrix (c_{ij}) into a unit matrix plus one extra column. We thus rewrite the system (1.1) in the form

$$\gamma_1 \mathbf{x}^{a_1} - \mathbf{x}^{a_{d+1}} \;=\; \gamma_2 \mathbf{x}^{a_2} - \mathbf{x}^{a_{d+1}} \;=\; \dots \;=\; \gamma_d \mathbf{x}^{a_d} - \mathbf{x}^{a_{d+1}} \;=\; 0, \qquad (3.1)$$

where the γ_i are $\mathbf{Q}$-linear combinations of the old coefficients c_{ij}. Equivalently, we have

$$\gamma_1 \mathbf{x}^{u_1 - a_{d+1}} \;=\; \gamma_2 \mathbf{x}^{a_2 \; a_{d+1}} \;=\; \dots \;=\; \gamma_d \mathbf{x}^{a_d - a_{d+1}} \;=\; 1. \qquad (3.2)$$

We now compute the *Smith normal form* of the $d \times d$-exponent matrix $(a_1 - a_{d+1}, \dots, a_d - a_{d+1})$. This means we construct invertible integer $d \times d$-matrices U and V such that

$$V \cdot \big(a_1 - a_{d+1}, a_2 - a_{d+1}, \dots, a_d - a_{d+1}\big) \cdot U \quad = \quad diag(m_1, m_2, \dots, m_d), \qquad (3.3)$$

where m_{i-1} divides m_i for all i. As in (2.4), we have $\mathbf{Z}^d / \mathbf{Z}\{\mathcal{A}\} \simeq \mathbf{Z}/m_1\mathbf{Z} \oplus \dots \oplus \mathbf{Z}/m_d\mathbf{Z}$, and $m_1 m_2 \cdots m_d = v(\mathcal{A})$, the normalized volume of the simplex $Q = conv(\mathcal{A})$.

The invertible matrix $U = (u_1, \dots, u_n)$ defines a monoidal transformation of coordinates $x_i \mapsto \mathbf{z}^{u_i}$. In the new coordinates $\mathbf{z} = (z_1, \dots, z_d)$ our system (3.3) equals

$$\tilde{\gamma}_1 z_1^{m_1} \;=\; \tilde{\gamma}_2 z_2^{m_2} \;=\; \dots \;=\; \tilde{\gamma}_d z_d^{m_d} \;=\; 1, \qquad (3.4)$$

where the $\tilde{\gamma}_i$ are obtained from the γ_j via the monoidal transformation defined by V.

By construction, the number of real roots of (1.1) equals the number of real roots of (3.4). This number is bounded above by $2^{\#\{m_i \, : \, m_i \, even\}}$. If $\nu = m_1 m_2 \cdots m_d$ is odd, then at least one of the roots is real. This completes our proof for the case $n = d + 1$.

For the general case we need the following description of the regular triangulation Δ_ω.

Lemma 3.1 *A subset* $\mathcal{B} \subset \mathcal{A}$ *is a cell of* Δ_ω *if and only if there exist* $\lambda_0, \lambda_1, \ldots, \lambda_d \in \mathbf{Q}$

$$\text{such that} \qquad \sum_{j=1}^{d} \lambda_j a_{ij} + \lambda_0 \quad \left\{ \begin{array}{ll} = \omega_i & \text{if } a_i \in \mathcal{B}, \\ > \omega_i & \text{if } a_i \notin \mathcal{B}. \end{array} \right. \tag{3.5}$$

Proof This follows from the equivalent definitions in Billera, Filliman and Sturmfels [1990], (4.4) and Lee [1991], Section 4. $\square$

Proof of Theorem 2.2 (Part II: $n > d+1$). We view the complex roots of (1.5) as the branches of a vector-valued algebraic function of t as $t \to 0$. The number of branches equals $v(\mathcal{A}) = \sum_{\mathcal{B} \in \Delta_\omega} v(\mathcal{B})$. For each cell $\mathcal{B}$ of Δ_ω we get $v(\mathcal{B})$ branches. These are computed using the following transformation.

We choose $\lambda_0, \lambda_1, \ldots, \lambda_d \in \mathbf{Q}$ as in Lemma 3.1. We substitute $x_i \cdot t^{-\lambda_i}$ for x_i and multiply each equation in (1.5) by $t^{-\lambda_0}$ to get the equivalent system

$$\frac{1}{t^{\lambda_0}} \cdot f_i\Big(t; \frac{x_1}{t^{\lambda_1}}, \ldots, \frac{x_d}{t^{\lambda_d}}\Big) \;\; = \;\; \sum_{a_j \in \mathcal{B}} c_{ij} \mathbf{x}^{a_j} + \sum_{a_\ell \notin \mathcal{B}} c_{i\ell} \, t^{\gamma_\ell} \, \mathbf{x}^{a_\ell}, \qquad (i = 1, 2, \ldots, d).$$
$$\tag{3.6}$$

The exponents γ_ℓ are positive rational numbers.

For $t = 0$ the system (3.6) has $v(\mathcal{B})$ complex roots. We may assume that there are no multiple roots, if necessary, by shrinking the Zariski open set $\mathcal{U}_\mathcal{A}$. Let $\rho_\mathcal{B}$ denote the number of real roots at $t = 0$. By the Implicit Function Theorem, the system (3.6) has $\rho_\mathcal{B}$ real roots for all parameters t in a small neighborhood of the origin 0.

We have shown that for $t \to 0$ the number of real roots of (1.5) equals

$$\rho \;\; = \;\; \sum_{\mathcal{B} \in \Delta_\omega} \rho_\mathcal{B}.$$

By part I each $\rho_\mathcal{B}$ satisfies the desired upper and lower bound. Since these bounds are additive with respect to the cells of Δ_ω, the proof of Theorem 2.2 is complete. $\square$

References

Bernstein, D.N. [1975] *The number of roots of a system of equations,* Functional Analysis and its Applications, **9 (2)**, 183–185.

Billera, L.J., Filliman, P., and Sturmfels, B. [1990] *Constructions and complexity of secondary polytopes,* Advances in Mathematics, **83**, 155–179.

Gel'fand, I.M., Kapranov, M.M., and Zelevinsky, A.V. [1990] *Discriminants of polynomials in several variables and triangulations of Newton polytopes,* Algebra i analiz, **2**, 1–62; English translation in *Leningrad Math. Journal,* **2** (1991).

Hungerford, T. [1974] *Algebra, Graduate Texts in Mathematics,* Springer, New York.

Khovanskii, A.G. [1980] *On a class of systems of transcendental equations,* Soviet Math. Doklady, **22 (3)**, 762–765.

———— [1991] *Fewnomials,* Translations of Math. Monographs, **88**, American Mathematical Society.

Kushnirenko, A.G. [1975] *The Newton polyhedron and the number of solutions of a system of k equations in k unknowns,* Uspekhi Mat. Nauk., **30**, 266–267.

Lee, C. [1991] *Regular triangulations of convex polytopes,* in Applied Geometry and Discrete Mathematics — The Victor Klee Festschrift, (P. Gritzmann and B. Sturmfels, eds.), American Math. Soc, DIMACS Series Vol. 4, Providence, R.I.

Pedersen, P., and Sturmfels, B. *Product formulas for resultants and Chow forms,* Mathematische Zeitschrift, 214 (1993) 377–396.

Sturmfels, B. [1991] *Gröbner bases of toric varieties,* Tôhoku Math. J., **43**, 249–261.

Fields Institute Communications
Volume **3**, 1994

Gradient Flows for Local Minima of Combinatorial Optimization Problems

Wing Shing Wong
Department of Information Engineering
The Chinese University of Hong Kong
Shatin, N.T.
Hong Kong

Abstract. Recently, Brockett and Wong reported a connection between gradient flows on $SO(n)$ and local search solutions for the Assignment Problem. A paper by Wong later extends this result to a class of *NP-hard* combinatorial optimization problems. This short note summarizes the results in the latter paper.

1 Introduction

The idea of performing numerical computation by continuous flows has received much attention in the past. For examples one can refer to Symes [1980], [1982], Deift, Nanda and Tomei [1983], Brockett [1991, 1989], Chu [1988], Bayer and Lagarias [1989], Bloch [1990], Brockett and Wong [1991], Faybusovich [1991a, b], and Watkins and Elsner [1990]. Karmarkar [1990], proposed using a steepest descent approach to solve combinatorial optimization problems, (see also Kamath, Karmarkar and Ramakrishnan [1990], Karmarkar, Ramakrishnan and Resende [1991], and Karmarkar and Thakur [1992]). In Brockett and Wong [1991], the gradient flow approach was applied to a class of combinatorial optimization problems known as the Assignment Problem. One of the key results reported in that paper is that for any given Assignment Problem, there is a simple correspondence between local minima of a certain local search algorithm and local minima of an associated gradient flow defined on $SO(n)$.

Since the Assignment Problem is known to have a polynomial time solution, an algorithm to find a local minimum does not have great significance. However, Wong, (to appear) showed that the gradient flow approach can be easily extended to a large class of combinatorial optimization problems that includes the Traveling Salesman Problem and the Graph Partitioning Problem. Since these problems are

---

1991 *Mathematics Subject Classification*. Primary 49M37;Secondary 34A34.

Supported in part by the Ministry of Colleges and Universities of Ontario and the Natural Sciences and Engineering Research Council of Canada while visiting The Fields Institute.

well-known to be *NP-hard*, local search algorithm is commonly used to approximately solve these problems. Thus, results connecting local search algorithms for these problems with gradient flows may have a more practical significance. Moreover, the problem of finding a local minimum may be computationally complex itself. The complexity issue of local search algorithms is an interesting topic first raised by Johnson, Papdadimitriou and Yannakakis [1988]. In particular, for the Graph Partitioning Problem with the *SWAP* neighborhood, (two partitions are neighbors if one can be made identical to the other by swapping two vertices) it was shown by Schäffer and Yannakakis [1991], that finding the local minimum from an arbitrary initial point is *NP-hard*. In this short note, we will summarize the results presented in Wong, (to appear).

2 Matrix representation of combinatorial optimization problems

Let $\mathcal{S}_n$ represent the *symmetric group* on n symbols. By an *n-symbol combinatorial optimization problem*, we mean the problem of finding a globally optimal value of an arbitrary function defined on $\mathcal{S}_n$. Without loss of generality, we deal with minimization problems only in this paper. Let $\mathcal{P}_n$ represent the set of n by n *permutation matrices*; there is a one-to-one, onto correspondence between elements in $\mathcal{S}_n$ and $\mathcal{P}_n$. By viewing $\mathcal{P}_n$ as an *incidence matrix* for a graph, one also obtains a one-to-one, onto correspondence between $\mathcal{P}_n$ and $\mathcal{G}_n$, the set of directed graphs with the property that every vertex is the source of exactly one directed edge and the sink of exactly one directed edge. (We use the convention that $P_{i,j}$ is equal to 1 if and only if there is a directed edge from vertex i to vertex j, i is called the source and j is the sink of the directed edge.)

Let $tr M$ represent the trace of matrix M. If C is the cost matrix defining an Assignment Problem, it is well-known that it can be represented in the form:

$$\min_{P \epsilon \mathcal{P}_n} \; tr \, C^T P. \tag{2.1}$$

It is possible to extend this representation to other combinatorial optimization problems. In fact, the following theorem holds:

Theorem 2.1 *Any* n-*symbol combinatorial optimization problem can be represented in the form:*

$$\min_{P \epsilon \mathcal{P}_n} \; tr \left(\sum_{i_1, i_2, \ldots, i_n} D_{i_1} P^T D_{i_2} P \cdots D_{i_n} P \right) \tag{2.2}$$

where the D_i's are n by n rank 1 matrices.

For the proof, see Wong, (to appear). For commonly encountered problems, the expression in Equation (2.2) can usually be simplified. We consider three classes of examples.

1. The Assignment Problem

It was shown in Brockett and Wong [1991], that the Assignment Problem can be defined in the form:

$$\min_{P \epsilon \mathcal{P}_n} tr \sum_{i=1}^{r} X_i P^T Y_i P \tag{2.3}$$

where r is the rank of the cost matrix, C, and X_i's and Y_i's are diagonal matrices obtained by the following *outer-product decomposition* of C: C:

$$C = \sum_{i=1}^{r} X_i \begin{bmatrix} 1 \\ \cdot \\ \cdot \\ \cdot \\ 1 \end{bmatrix} [1, \cdots, 1] \, Y_i. \tag{2.4}$$

2. The Traveling Salesman Problem

The Traveling Salesman Problem is similar to the Assignment Problem except that the optimization is restricted to *n-cycles* only. In the graphical representation, these permutations correspond to the *directed circuits with n-vertices*. Although, strictly speaking, this is not a problem defined on $\mathcal{S}_n$, we can treat the Traveling Salesman Problem as an n-symbol combinatorial optimization problem by making use of the observation that the set of permutation matrices corresponding to all n-cycles is equal to the set of the matrices of the form $P^T S_{\mathcal{T}}(n) P$, where P is a permutation matrix and

$$S_{\mathcal{T}}(n) = \begin{bmatrix} 0, & 1, & 0, & \cdots, & 0 \\ 0, & 0, & 1, & \cdots, & 0 \\ \cdot & \cdot & \cdot & \cdots, & \cdot \\ \cdot & \cdot & \cdot & \cdots, & \cdot \\ 0, & 0, & 0, & \cdots, & 1 \\ 1, & 0, & 0, & \cdots, & 0 \end{bmatrix}.$$

Hence, the Traveling Salesman Problem can be formulated as:

$$\min_{P \epsilon \mathcal{P}_n} tr \, C^T P^T S_{\mathcal{T}}(n) P. \tag{2.5}$$

Notice that, in this formulation, different elements in $\mathcal{S}_n$ may correspond to the same directed circuit.

3. The Graph Partitioning Problem

Let G be a fully connected undirected graph with n vertices with weights assigned to its undirected edges. Let p and q be two positive integers such that their sum is equal to n. The Generalized Graph Partitioning Problem is to find a partition of vertices into two subsets with p and q elements such that the sum of the weights on the cut edges (that is, edges with their endpoints in different subsets of the partition) is minimized. The case where p and q are equal defines the classic Graph Partitioning Problem.

Just like the Traveling Salesman Problem, the Graph Partitioning Problem is not defined on $\mathcal{S}_n$, strictly speaking. However, we can formulate a corresponding problem on $\mathcal{S}_n$ by first constructing a fully connected directed graph with n vertices.

For any original weight assigned to an undirected edge, we assign half of it to each of the two directed edges joining the same vertices in the undirected graph. Call the cost matrix corresponding to this new set of weights, C. Now define

$$S_{\mathcal{G}}(n) = \begin{bmatrix} 0_{p,p}, & 1_{p,q} \\ \\ 1_{q,p}, & 0_{q,q} \end{bmatrix} \tag{2.6}$$

where $1_{i,j}$ denotes the i by j matrix with all entries equal to 1.

Then, the Graph Partitioning Problem can be represented in the following way:

$$\min_{P \epsilon \mathcal{P}_n} \; tr \, C^T P^T S_{\mathcal{G}}(n) P. \tag{2.7}$$

3 Local search and 2-change neighborhood

For the rest of this paper, we will consider combinatorial optimization problems of the form:

$$\min_{P \epsilon \mathcal{P}_n} \; tr \, \sum_{i=1}^{r} X_i^T P^T Y_i P \tag{3.1}$$

where X_i and Y_i are arbitrary matrices. Thus, this class of problems contains all three examples presented in the previous section.

Local search algorithms are commonly used to approximately solve computational hard combinatorial optimization problems. All these algorithms require the definition of a neighborhood around an arbitrary element in the search space. The *k-change neighborhood* is a popular concept that is also known to be quite efficient for the Traveling Salesman Problems and the Graph Partitioning Problems.

For any n-symbol combinatorial optimization problem, we can define a *k-change neighborhood* in the following way. For any matrix P in $\mathcal{P}_n$, let G be the corresponding directed graph in $\mathcal{G}_n$. The *k-change neighborhood* of G (and hence P), is then defined as the set of elements in $\mathcal{G}_n$ that can be obtained by removing k directed edges from G and then placing k alternative directed edges to the remaining graph. As an example, consider the permutation matrix

$$\begin{bmatrix} 0, & 1, & 0, & 0 \\ 0, & 0, & 1, & 0 \\ 0, & 0, & 0, & 1 \\ 1, & 0, & 0, & 0 \end{bmatrix}.$$

Its six neighbors in the *2-change neighborhood* are represented by the matrices:

$$\begin{bmatrix} 0, & 1, & 0, & 0 \\ 1, & 0, & 0, & 0 \\ 0, & 0, & 0, & 1 \\ 0, & 0, & 1, & 0 \end{bmatrix}, \begin{bmatrix} 0, & 0, & 0, & 1 \\ 0, & 0, & 1, & 0 \\ 0, & 1, & 0, & 0 \\ 1, & 0, & 0, & 0 \end{bmatrix}, \begin{bmatrix} 0, & 1, & 0, & 0 \\ 0, & 0, & 1, & 0 \\ 1, & 0, & 0, & 0 \\ 0, & 0, & 0, & 1 \end{bmatrix},$$

$$\begin{bmatrix} 1, & 0, & 0, & 0 \\ 0, & 0, & 1, & 0 \\ 0, & 0, & 0, & 1 \\ 0, & 1, & 0, & 0 \end{bmatrix}, \quad \begin{bmatrix} 0, & 0, & 1, & 0 \\ 0, & 1, & 0, & 0 \\ 0, & 0, & 0, & 1 \\ 1, & 0, & 0, & 0 \end{bmatrix}, \quad \begin{bmatrix} 0, & 1, & 0, & 0 \\ 0, & 0, & 0, & 1 \\ 0, & 0, & 1, & 0 \\ 1, & 0, & 0, & 0 \end{bmatrix}.$$

If P is an element in $\mathcal{P}_n$, it is easy to see that the 2-change neighborhood of P consists of the $n(n-1)/2$ permutation matrices of the form $PN_{i,j}$, where $N_{i,j}$ is a permutation matrix that has only two non-zero off diagonal entries, at (i,j) and (j,i). With this as motivation, we formally define:

Definition 3.1 *For the combinatorial optimization problem defined in (3.1), the 2-change neighborhood of a permutation matrix P consists of P and matrices of the form:*

$$PN_{i,j}$$

for $1 < i < j < n$. An element of the form $PN_{i,j}$ where

$$(PN_{i,j})^T Y_k PN_{i,j} \neq Y_k$$

for some k, $(1 \leq k \leq r)$, is called a neighbor of P.

Local search algorithms come in many flavors. Since these details will not affect our discussion, we will simply assume that a fixed local search algorithm has been chosen from now on.

Definition 3.2 *A 2-opt solution for the combinatorial optimization problem defined in (3.1) is defined as a locally minimal solution obtained under the chosen local search algorithm using the 2-change neighborhood. The solution is called* non-degenerate *if its value is strictly lower than the value at its neighbors.*

Notice that the definition of a 2-change neighborhood is independent of the parameters of the problem, but the definition of a distinct neighbor is dependent on the Y_k's. The motivation for this will become clear when we study the Graph Partitioning Problem and the Traveling Salesman Problem.

To illustrate the definition of a 2-change neighborhood, let us apply it to a Graph Partitioning Problem with $p = q = 2$. Recall that the cost function for this problem is formulated as: $tr\, C^T P^T S_{\mathcal{G}}(n)P$ with the state of the partition represented by $P^T S_{\mathcal{G}}(n)P$. For simplicity, consider the neighborhood around the identity matrix, I_n. The matrix,

$$I^T S_{\mathcal{G}}(n)I = \begin{bmatrix} 0, & 0, & 1, & 1 \\ 0, & 0, & 1, & 1 \\ 1, & 1, & 0, & 0 \\ 1, & 1, & 0, & 0 \end{bmatrix}$$

represent the state that divides the graph into two subgraphs, one consisting of vertices 1 and 2, the other consisting of vertices 3 and 4. The matrices $N_{1,3}^T S_{\mathcal{G}}(n)N_{1,3}$ is given by:

$$\begin{bmatrix} 0, & 1, & 1, & 0 \\ 1, & 0, & 0, & 1 \\ 1, & 0, & 0, & 1 \\ 0, & 1, & 1, & 0 \end{bmatrix}.$$

This corresponds to the partition that groups vertex 1 with 4 and vertex 2 with 3 (we have swapped vertex 1 with vertex 3). Since vertex 1 and vertex 2 are in the same subset in the original partition, swapping the two vertices should produce to the same partition. This is confirmed by the fact that $N_{1,2}^T S_G N_{1,2} = S_G$. Moreover, according to our definition, $N_{1,2}$ is not a distinct neighbor of the identity matrix. Generalizing this argument, one can show that this definition of the 2-change neighborhood coincides with the definition of the $SWAP$ neighborhood. Moreover, a non-degenerate 2-opt solution defined here is a non-degenerate 2-opt solution in the classical sense.

Applying this concept to the Traveling Salesman Problem leads to two types of new circuits as illustrated in Figures 1 and 2. It is clear that this definition of the neighborhood is related but not identical to the classical definition of a 2-opt neighborhood for the Traveling Salesman Problem. The difference comes from the fact that the neighborhood defined here is obtained by interchanging vertices rather than edges.

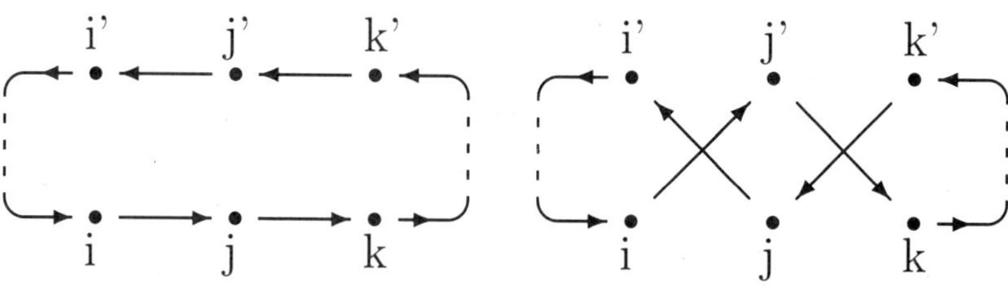

<table>
<tr><td align="center">Directed Circuit Corresponding to Θ</td><td>Directed Circuit Corresponding to $\Theta N_{j,j'}$ when Vertex j and Vertex j' are not directly connected.</td></tr>
</table>

Figure 1

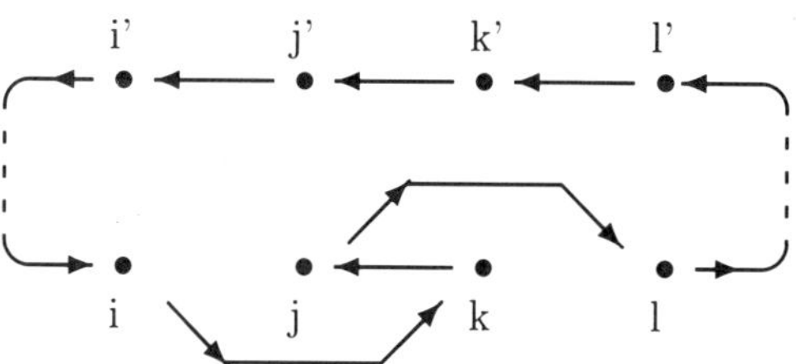

Directed Circuit Corresponding to $\Theta N_{j,k}$ when there is an edge from vertex j to vertex k.

Figure 2

4 Embedding in $\mathcal{SO}(n)$ and equivalency of optimality

Since $\mathcal{P}_n$ is a subset of $\mathcal{O}(n)$, (a subgroup in fact) one can embed the optimization problem defined in (3.1) to an optimization problem defined in $\mathcal{O}(n)$, namely:

$$\min_{\Theta \epsilon \mathcal{O}(n)} tr \sum_{i=1}^{r} X_i^T \Theta^T Y_i \Theta. \tag{4.1}$$

For the case where X_i's and Y_i's are diagonal matrices, this type of optimization problem, restricted to the connected component containing the identity matrix, $\mathcal{SO}(n)$, was considered in detail in Brockett [1991] and was shown to be related to solving geometric matching problems.

For cases where the X_i's and Y_i's are not all diagonal, the optimization problem defined in (4.1) has one undesirable feature. Namely, since $\mathcal{O}(n)$ contains elements of the form DP, (the so-called *hyperocathedral matrices*) where D is a diagonal matrix with diagonal values 1 or -1, this embedding may introduce many spurious local minima. In order to eliminate these spurious local minima, one can define a different embedded problem by defining a cost function η on $\mathcal{O}(n)$ by

$$\eta(\Theta) = tr \sum_{i=1}^{r} X_i^T (\Theta \circ \Theta)^T Y_i \Theta \circ \Theta, \tag{4.2}$$

where $M \circ N$ denote the Schur-Hadamard product of two matrices defined by $(M_{i,j} N_{i,j})$. If we identify elements in $\mathcal{P}_n$ with elements in $\mathcal{SO}(n)$ by the following mapping:

$$\iota(P) = \begin{cases} P & \quad if\ det(P) = 1 \\ \begin{bmatrix} -1 & 0 & \cdots & 0 \\ 0 & 1 & \cdots & 0 \\ . & . & \cdots & . \\ 0 & 0 & \cdots & 1 \end{bmatrix} P & \quad if\ det(P) = -1. \end{cases}$$

Then, it is trivial to see that:

$$\eta(\iota(P)) = tr \sum_{i=1}^{r} X_i^T P^T Y_i P. \tag{4.3}$$

So, the values of the cost function at all permutation matrices are preserved and it is sufficient to consider the optimization problem on $\mathcal{SO}(n)$.

So, from now on, we will consider an embedded optimization problem of the form:

$$\min_{\Theta \epsilon \mathcal{SO}(n)} tr\ \eta(\Theta). \tag{4.4}$$

Since the inverse of ι is given by:

$$\iota^{-1}(H) = H \circ H$$

we can extend the definition of ι^{-1} to all hyperoctahedral matrices. Notice that different local minima of η may correspond to the same element in $\mathcal{P}_n$.

Although the value of the cost function is preserved in this setting, there still remains the important question of whether an optimal solution in the original combinatorial optimization problem is preserved as a global optimum for the problem defined in (4.4). The answer to this question is positive for the Least Square Matching Problem and the Assignment Problem. It is not known to hold for a problem defined in (3.1) in general. A more realistic requirement is that the local optimality structure be preserved.

Since η can be viewed as a potential function on a compact Riemannian manifold, it is natural to consider the gradient flow it induces. It is well-known that $\mathcal{SO}(n)$ is a Lie group and the tangent space at any of its element can be identified with the space of anti-symmetric matrices, $o(n)$. Define an inner-product on $o(n)$ by:

$$< \Pi, \Omega >= tr(\Pi^T \Omega).$$

This inner-product defines a Riemannian structure on $\mathcal{SO}(n)$.

It is natural to study the gradient flow induced by η under this standard Riemannian structure. Moreover, if the embedding defined previously preserves the local optimality structure, we may expect that there should be a correspondence between asymptotically stable critical points of this induced gradient flow and 2-opt solutions. The main result of this paper is to prove that this correspondence holds for certain combinatorial optimization problems.

Denote the gradient flow induced by η by:

$$\frac{d\Theta}{dt} = -\nabla\eta(\Theta(t)). \tag{4.5}$$

Strictly-speaking, Equation (4.5) should be interpretated as a dynamical equation on $\mathcal{SO}(n)$. However, since $\mathcal{SO}(n)$ is embedded in the space of all n by n matrices, one can view Equation (4.5) as a matrix differential equation with the understanding that if the system starts at a point in $\mathcal{SO}(n)$, its trajectory for all time remains in $\mathcal{SO}(n)$. As observed by L. Fayusovich, this equation can be numerically integrated without requiring any matrix inversion, unlike many other continuous flow algorithms.

Definition 4.1 *The weak gradient flow equivalency property is said to hold for a combinatorial optimization problem defined in (3.1) if*

1. *The image of every non-degenerate 2-opt solution under ι is an asymptotically stable critical point for the gradient flow system defined by Equation (4.5).*

2. *For every H, a hyperoctahedral, asymptotically stable critical point of the gradient flow system defined by Equation (4.5), $\iota^{-1}(H)$ is a 2-opt solution.*

We call this a weak equivalency because the second condition is a weak converse to the first condition. It is weak for two reasons:

1) It requires the assumption that the asymptotically stable critical point is hyperoctahedral, and

2) It does not require the corresponding 2-opt solution be non-degenerate.

We will show that for the Graph Partitioning Problem and some cases of the Traveling Salesman Problem, the weak gradient flow equivalency property holds.

5 Conditions for weak gradient equivalency

For any $1 \leq i < j \leq n$, define the matrix $\Omega(i,j)$ by:

$$\Omega(i,j)_{p,q} = \begin{cases} 1, & if\ i = p, j = q \\ -1, & if\ i = q, j = p \\ 0, & otherwise. \end{cases}$$

The $\Omega(i,j)$'s form a basis for $o(n)$. Define

$$\Lambda(i,j)_{p,q} = \begin{cases} 1, & if\ p = q = i \\ 1, & if\ p = q = j \\ 0, & otherwise \end{cases}$$

and

$$\Pi(i,j) = \Omega(i,j) \circ \Omega(i,j) - \Lambda(i,j).$$

So, $\Pi(i,j)$ is the symmetric matrix with -1 at the (i,i) and (j,j) entries and 1 at the (i,j) and (j,i) entries, and zero elsewhere.

Theorem 5.1 *(Wong, (to appear)) The weak gradient flow equivalency property holds for the combinatorial optimization problem defined in (3.1) if and only if at any hyperoctahedral critical point, Θ, of Equation (4.5), the following property holds for all $1 \leq i < j \leq n$:*

If $(\eta(\Theta N_{i,j}) - \eta(\Theta)) > 0$, then
$$\left(\eta(\Theta N_{i,j}) - \eta(\Theta) - tr\ \textstyle\sum_{k=1}^{r} X_k^T \left(\Theta \Pi(i,j)\right)^T Y_k \Theta \Pi(i,j) \right) > 0.$$

If $\left(\eta(\Theta N_{i,j}) - \eta(\Theta) - tr\ \sum_{k=1}^{r} X_k^T \left(\Theta \Pi(i,j)\right)^T Y_k \Theta \Pi(i,j) \right) \geq 0$, then (5.1)
$(\eta(\Theta N_{i,j}) - \eta(\Theta)) \geq 0.$

For the proof of this theorem, please refer to Wong, (to appear).

At a first glance, the condition in Theorem 5.1 may be very difficult to verify. However, observe that if $tr\ \sum_{k=1}^{r} X_k^T \Pi(i,j) \widetilde{Y_k}(\Theta) \Pi(i,j) = 0$, then conditions in (5.1) hold automatically. For any n by n matrix M, define

$$\Delta_{i,j}(M) = M_{i,j} + M_{j,i} - M_{i,i} - M_{j,j}. \tag{5.2}$$

Then it is straight-forward to check that

$$tr X_k^T \Pi(i,j)\widetilde{Y_k}(\Theta)\Pi(i,j) = \Delta_{i,j}(X_k)\Delta_{i,j}(\widetilde{Y_k}(\Theta)). \tag{5.3}$$

Moreover, the following result holds:

Theorem 5.2 *(Wong, (to appear)) If for any $1 \leq k \leq r$, $\Delta_{i,j}(X_k) = 0$ for all i,j, or $\Delta_{i,j}(Y_k) = 0$ for all i,j, then the weak gradient equivalency property holds for the problem defined by (3.1).*

Define a matrix, M, to be *locally balanced* if $\Delta_{i,j}(M) = 0$ for all i,j. It is easy to show that a matrix is locally balanced if and only if it is of the form

$$A + \begin{bmatrix} 1 \\ \vdots \\ 1 \end{bmatrix} c^T \tag{5.4}$$

where A is anti-symmetric and and c is an arbitrary n-dimension row vector.

In the next section, we will present specific examples where the conditions of Theorem 5.2 are satisfied.

6 Weak gradient equivalency property of the GPP and TSP

We have previously discussed how to represent the Graph Partitioning Problem (GPP) as a problem on $\mathcal{P}_n$. Given a GPP, there is no loss in generality in assuming that all the diagonal elements of the cost matrix are zero and all its entries are non-negative. For such a modified problem, the following result holds:

Theorem 6.1 *(Wong, (to appear)) All Graph Partitioning Problems have the weak gradient equivalency property.*

For the Traveling Salesman Problem (TSP) with a locally balanced cost matrix, it follows from Theorem 5.2 that such a problem also has the weak gradient equivalency property. However, for an arbitrary cost matrix C, the weak gradient equivalency property may not hold.

References

Bayer, D.A., and Lagarias, J. [1989] *The Nonlinear Geometry of Linear Programming I and II*, Trans. Amer. Math. Soc., **314**, 499–526, 527–581.

Bloch, A.M. [1990] *Steepest Descent, Linear Programming and Hamiltonian Flows*, Contemporary Math, AMS, **114**, 77–88.

Brockett, R.W. [1991] *Dynamical Systems that Sorts and Solve Linear Programming Problems;* in 'Proc. 27th IEEE Conf. on Decision and Control', IEEE, 1988, 799–803, subsequently published in Linear Algebra and Its Applications, **146**, 79–91.

———— [1989] *Least Squares Matching Problems*, Linear Algebra and Its Applications, **122**, 761–777.

Brockett, R.W., and Wong, W.S. [1991] *A Gradient Flow for the Assignment Problem;* in Progress in Systems Theory (G. Conte and B. Wyman, eds.), 170–177.

Chu, M.T. [1988] *On the Continuous Realization of Iterative Processes*, SIAM Review, **30** (3), 375–389.

Deift, T., Nanda, and Tomei, C. [1983] *Differential equations for the symmetric eigenvalue problem*, SIAM J. on Numerical Analysis, **20**, 1–22.

Faybusovich, L. [1991a] *Hamiltonian structure of dynamical systems which solve linear programming problems*, Physica D, **53**, 217–232.

———— [1991b] *Dynamical systems which solve optimization problems with linear constraints*, IMA J. of Mathematical Control & Information, **8**, 135–149.

Johnson, D.S., Papdadimitriou, C.H., and Yannakakis, M. [1988] *How Easy is Local Search?* J. Comput. System Sci., **37**, 79–100.

Kamath, A.P., Karmarkar, N.K., Ramakrishnan, K., and Resende, M.G.C. [1990] *Computational Experience with an Interior Point Algorithm on the Satisfiability Problems*, Annals of Operations Research, **25**, 43–58.

Karmarkar, N. [1990] *An Interior-Point Approach to NP-Complete Problems - Part 1*; in Mathematical Developments Arising From Linear Programming (J.C. Lagarias and M.J. Todd, eds.), Comtemporary Math., **114**, 297–308.

Karmarkar N.K., Ramakrishnan, K., and Resende, M.G.C. [1991] *An Interior-Point Algorithm to Solve Computationally Difficult Set Covering Problems*, Math. Prog., Series B, **52**, 597–618.

Karmarkar, N.K., and Thakur, S.A. [1992] *An Interior-Point Approach to Tensor Optimization Problem with Application to Upper Bounds in Integer Quadratic Optimization Problems*, Proc. 2nd Conference on Integer Programming and Combinatorial Optimization, 406–420.

Schäffer, A., and Yannakakis, M. [1991] *Simple Local Search Problems that are Hard to Solve*, SIAM J. Comput., **20**(1), 56–87.

Symes, W.W. [1980] *Hamiltonian Group Actions and Integrable Systems*, Physica D, **1**, 339–376.

———— [1982] *The QR Algorithm and Scattering for the Nonperiodic Toda Lattice*, Physica D, **4**, 275–280.

Watkins, D.S. and Elsner, L. [1990] *On Rutishauser's Approach to Self-Similar Flows*, SIAM J. Matrix Anal. Appl., **11** (2), 301–311.

Wong, W.S. *Matrix Representation and Gradient Flows for NP-Hard Problems*, (to appear).